Utilizing Information Technology in Developing Strategic Alliances Among Organizations

Salustiano Martínez-Fierro, University of Cádiz, Spain

José Aurelio Medina-Garrido, University of Cádiz, Spain

Jose Ruiz-Navarro, University of Cádiz, Spain

IDEA GROUP PUBLISHING

Hershey • London • Melbourne • Singapore

Acquisitions Editor: Michelle Potter
Development Editor: Kristin Roth
Senior Managing Editor: Amanda Appicello
Managing Editor: Jennifer Neidig
Copy Editor: Holly Powell
Typesetter: Marko Primorac
Cover Design: Lisa Tosheff
Printed at: Yurchak Printing Inc.

Published in the United States of America by
Idea Group Publishing (an imprint of Idea Group Inc.)
701 E. Chocolate Avenue
Hershey PA 17033
Tel: 717-533-8845
Fax: 717-533-8661
E-mail: cust@idea-group.com
Web site: http://www.idea-group.com

and in the United Kingdom by
Idea Group Publishing (an imprint of Idea Group Inc.)
3 Henrietta Street
Covent Garden
London WC2E 8LU
Tel: 44 20 7240 0856
Fax: 44 20 7379 0609
Web site: http://www.eurospanonline.com

Library of Congress Cataloging-in-Publication Data

Utilizing information technology in developing strategic alliances among organizations / Salustiano Martinez-Fierro ... [et al.], editor.
p. cm.
Summary: "The book analyzes the development of global business-to-business electronic markets, and whether these markets are becoming a way of improving trust between organizations"--Provided by publisher.
ISBN 1-59140-893-8 (hardcover) -- ISBN 1-59140-894-6 (softcover) -- ISBN 1-59140-895-4 (ebook)
1. Strategic alliances (Business) 2. Information technology--Management. I. Martinez-Fierro, Salustiano, 1973-
HD69.S8U85 2006
658'.046--dc22

2006009286

British Cataloguing in Publication Data
A Cataloguing in Publication record for this book is available from the British Library.

Utilizing Information Technology in Developing Strategic Alliances Among Organizations

Table of Contents

Preface

Strategic Importance of Alliances and Support Provided by Information Technology

Research in strategic management has interpreted how companies compete in very diverse ways. Researchers have elaborated different theories or approaches, whose reasoning ranges from explanations oriented towards the interior of the firm to explanations taking the external environment as the point of reference (Hoskisson, Hitt, Wan, & Yiu, 1999). However, the majority of these approaches regard the firm as an independent and autonomous entity that competes with other firms for returns in an impersonal market. But affirming this, categorically, is inappropriate in a world in which firms belong to networks of social, professional, and exchange relationships with other companies (Galaskiewicz & Zaheer, 1999; Granovetter, 1985; Gulati, 1998). These networks consist of lasting ties that have strategic significance for the firms belonging to them. They include strategic alliances, joint ventures, long-term collaboration relationships between suppliers and their customers, and a great number of similar ties (Gulati, Nohria, & Zaheer, 2000).

As Osborn and Hagedoorn (1997) affirm, researchers are beginning to realize that the modern world is dominated by organizations with specialized skills. It is, consequently, also a world dominated by the connections between these specialized organizations.

Neither Market nor Hierarchy: Networks and Alliances

Franchises, such as those created by the pioneering Kentucky Fried Chicken or McDonald's, in which the use of a brand, image, and commercial formula is licensed to third parties unconnected with the ownership of the franchisor, did not stop growing throughout the second half of the 20th century. The same can be said of the close vertical relationships arising within firms from the same industry, such as in Benetton or Toyota.

For Jarillo (1993), the previous cases are examples of successful firms that do two apparently contradictory things at the same time. First, these companies control the entire production process, from the raw materials to the end consumer. Second, they do not own the units that supply them with their raw materials or subcomponents, nor the units commercializing the end products with the end consumers. In other words, these companies act simultaneously as large integrated firms, taking care of everything, and as firms concentrating only on a few key activities and outsourcing the rest.

Most traditional organizations concentrate on only one of these two things, either carrying out most of the activities internally (the firms are then said to be *integrated* or *hierarchical* organizations), or alternatively delegating the supply of components or the distribution of end products to a number of arm's-length subcontractors for an agreed price. This second group of firms has no influence whatsoever over the subcontractors, and the relationship does not go beyond the contractually agreed transaction (hence, adopting so-called *market governance* forms). However, being able to operate in both ways simultaneously—combining the control capacity of the integrated firm with the flexibility and cost competence of market governance—is sometimes a requisite for success.

Best (1990) affirms that the large hierarchical firm is the old way of organizing things, while the new way of competing is based on a network of lateral and horizontal interconnections within companies and between them. According to Jarillo (1988, 1993), if a company is able to externalize certain non-core activities to the most efficient suppliers without its transaction costs increasing excessively, and undertakes the activities in which it has a comparative advantage itself, a superior form of organization emerges: the strategic network.

Networks and strategic alliances represent an attempt to resolve the competitive agility paradox (Osborn, 1998). According to this paradox, organizations need to be stable in order to be able to learn and grow by developing their strengths; at the same time, they need to be flexible in order to face a hypercompetitive environment (D'Aveni, 1994). Discontinuities in the environment make it impossible for an integrated firm to undertake all activities within its organization (Limerick & Cunnington, 1993). Thus, the dichotomy offered by Williamson (1975, 1985) in his initial work, regarding the choice between vertically integrated organizational forms and organizations carrying out free exchanges in the market, proves incapable of resolving this paradox.

Alliances and network forms represent an intermediate situation between the hierarchical governance forms typical of integrated firms and the market governance forms (Williamson, 1991), or as Thorelli (1986) points out, they are a distinct alternative from these two forms. According to Zenger and Hesterly (1997), a series of fundamental changes have dramatically altered the way in which production and services are organized in the economy: firms have shrunk in size; the exchanges occurring inside companies are becoming more like those of the market; and the governance mechanisms typical of the hierarchical firm are increasingly assisting transactions between companies operating in the market. This gives rise to network organizational forms consisting of small autonomous units that possess elements of both forms of governance—hierarchy and market—and which may also have started out either as one form or the other. In this respect, integrated companies fragment into autonomous units that are appropriately coordinated, and those organizations that once operated in the free market are now coordinated in networks.

At this point it might be interesting to point out that alliances are a concept that is not exclusive to modern times, nor are they inconceivable without the support of information technology (IT), which we shall discuss in the following section. An example of an alliance was founded by 20 Greek cities to defeat Persia in around 448 BC (Smith, Carroll, & Ashford, 1995). However, IT undoubtedly generates the driving force that allows alliances to expand rapidly (García Calderón Díaz, 1998). Currently, alliances are something more than just an option: the dynamism of the environment, nowadays, is creating a context in which these alliances are essential to resolve the previously described paradox, and alliances are proliferating in large part due to the support modern IT provides in solving the problems that arise in their information flows.

Alliances, Information, and Information Technology

Alliances can provide a company with access to information, resources, markets, technologies, facilities for learning, and economies of scale and scope; they also allow firms to share risks and subcontract some of their value-chain activities (Gulati et al., 2000).

Various factors enable network organizational forms to develop. Specifically, technological and organizational innovations facilitate the implementation of market mechanisms in hierarchies, making it possible for firms to structure themselves into smaller units, measure performance, and reward achievement accordingly. Zenger and Marshall (1995) argue that the smaller the units, the more effective the incentives. Moreover, these technological and organizational innovations have accelerated the implementation of hierarchy elements within markets, including control, the administration of incentives, orders, and a richer communication. In this context, IT plays an important role, whatever the origin of the alliances or networks created. IT allows these disaggregated units to be monitored and their performance to be measured; likewise, it improves communication between the participating units and their coordination (Zenger & Hesterly, 1997).

According to Thorelli (1986), it may be useful to regard the alliance as a single organization with strategic goals. For this author, considering the functional proximity between the alliance members, it is also natural to take this perspective in the information systems used for the management and planning. He has no doubt about this, arguing that members that belong to an alliance will expect a large amount of information to be shared, so the majority of alliances between firms will need to be founded on their respective information systems.

Similarly, Limerick and Cunnington (1993) also stress the importance of IT. For these authors, IT is the driving force behind the expansion of strategic alliances. The new ITs have made it possible to design a set of more disaggregated, distributed, and flexible production and organization agreements, as well as, new forms in which companies can organize their internal operations and link up with the firms with which they are transacting (García Calderón Díaz, 1998).

Although many works have examined the topic of networks or relationships between companies,[1] and some have mentioned the importance of IT in them (Cash & Konsynski, 1985; Jarillo, 1993; Porter & Millar, 1986; Rackoff, Wiseman, & Ullrich, 1985), few contributions have explained IT's role in alliance formation in detail (see Clemons & Row, 1992; Kumar & van Dissel, 1996).

The literature has some gaps with regards to what IT does and how it enables—or at least improves—the functioning of alliances. In this respect, this book aims to take this opportunity and shed some light on the question.

Organization of This Book

This book deals jointly with two important topics for a firm's competitiveness in the framework of the information society: alliance strategies and the role played by IT in alliances. After reviewing the main concepts, theories, and work dealing with alliances, we analyze why firms form alliances and how IT influences the process. Besides this, the book analyzes: the development of global business-to-business electronic markets and whether these markets are becoming a way of improving trust between organizations at an international level; problems in implementing inter-organizational systems and their solutions; the importance of strategic alliances in building collaborative advantages by sharing knowledge or cooperating in innovation development; the relationships between the structures of inter-organizational information systems (IOS) and strategic alliances; how IT and alliance strategy are important both to multinational corporations and family firms, regardless of firm size; and how IT and cooperation support environmental management innovations through virtual cooperation between firms and institutions.

The book is organized into 11 chapters. A brief description of each of these chapters follows:

Chapter I defines some basic concepts about alliances; their attributes, the differences between them and traditional structures; and the types of alliance. The second section of this chapter analyzes the state of the art in the literature on alliances. The third and final section analyzes the role of IT in the formation and running of alliances.

Chapter II addresses the disorder that is apparent in the literature on inter-organizational relationships. This makes this field ideal for conducting quantitative studies that clarify the conceptual map and complement the existing qualitative research. In this chapter, a bibliometric study reveals the theoretical fragmentation of the literature on inter-organizational relationships; maps the most relevant studies and theoretical approaches; and highlights the topics that have been addressed most in this field.

Chapter III analyzes the development of global business-to-business electronic markets, and whether these markets are becoming a way of improving trust between organizations at an international level, thereby increasing the inter-organizational cooperation between them. This chapter proposes a model to explain the trust-related sources of competitive advantage for the new intermediaries in these markets and to compare these with off-line markets' characteristics. In addition, some propositions related with the trust-building mechanisms are defined.

Chapter IV stresses that while technical solutions to distance work abound, a strong and theoretically grounded approach on how best to implement these solutions is lacking. This chapter presents a three-part framework for organizing research in the areas of shared cognition, providing strategic alliance managers with a resource for implementing their technologically-mediated teams.

Chapter V establishes that alliances can be a source for firms to acquire and improve their knowledge-based capabilities. This chapter also analyzes the role of IT in the process of transmission and sharing of knowledge. The authors contend that alliances can also be a mechanism for firms to build a competitive advantage because of their proven access to economies of scope and scale, and complementary capabilities and knowledge.

Chapter VI analyzes the repercussions of the adoption of information and communication technology (ICT) on collaborative innovation. The authors aim to shed some light on the subject by means of an empirical analysis conducted on a sample of 1,649 Spanish manufacturing firms.

Chapter VII analyzes the relationships between the structures of IOS and strategic alliances. The authors contend that the relationship between both structures tends to be framed within a collaborative dynamic in which the IOS serves the objectives of the alliance. However, the authors also argue that there may be a competitive relationship between the IOS and the strategic alliance when both structures compete for firms from the same sector and geographical area.

Chapter VIII analyzes the potential impact of IT on the internationalization processes of family firms via cooperation agreements. The authors address the main research questions concerning the management of family firms and internationalization processes via strategic alliances. Subsequently, they carry out an analysis to determine IT's impact on family firms' internationalization processes.

Chapter IX focuses on the merger between two GSM operators in Turkey. This chapter is concerned with the structural changes in the IT department, and presents a detailed analysis of the different aspects of the merger process, as well as comments based on a review of the literature on information and communication technologies.

Chapter X analyzes the stock market reaction to the internationalization and diversification of the European telecom firms through acquisitions or strategic alliances. The authors analyze these operations with the purpose of verifying to what extent the stock market reaction to these business combinations is influenced by information asymmetry, resource complementarity, and management costs.

Chapter XI concludes and focuses on a particular type of technological relationship between firms and institutions for developing environmental management innovations through virtual cooperation. It goes into particular depth on the economic and strategic reasons explaining the formation of alliances based on IT. The author presents an example of a regional initiative for a cooperative platform based on IT that can result in the assimilation of technological and organizational competences in environmental management by firms located in the area.

References

Best, M. (1990). *The new competition.* Cambridge, MA: Harvard University Press.

Cash, J. I., & Konsynski, B. R. (1985). IS redraws competitive boundaries. *Harvard Business Review, 63*(2), 134-142.

Clemons, E. K., & Row, M. C. (1992). Information technology and industrial cooperation: The changing economics of coordination and ownership. *Journal of Management Information Systems, 9*(2), 9-28.

Contractor, F. J., & Lorange, P. (1988). *Cooperative strategies in international business.* Lexington: Lexington Books.

D'Aveni, R. (1994). *Hypercompetition: Managing the dynamics of strategic maneuvering.* New York: Free Press.

Galaskiewicz, J., & Zaheer, A. (1999). Networks of competitive advantage. In S. Andrews & D. Knoke (Eds.), *Research in the sociology of organizations* (pp.237-261). Greenwich, CT: JAI Press.

García Calderón Díaz, L. (1998). *A PLS model of strategic networks and organizational performance: Based on environmental turbulence, information technology, and transaction costs.* Doctoral Thesis. Instituto Tecnológico y de Estudios Superiores de Monterrey.

García Falcón, J. M., & Medina Muñoz, D. R. (1998). Enfoques teóricos sobre las relaciones interorganizativas: Una revisión comparativa. *Revista Europea de Dirección y Economía de la Empresa, 7*(3), 9-30.

Granovetter, M. S. (1985). Economic action and social structure: The problem of embeddedness. *American Journal of Sociology, 91,* 481-510.

Gulati, R. (1995). Does familiarity breed trust? The implications of repeated ties for contractual choice in alliances. *Academy of Management Journal, 38,* 85-112.

Gulati, R. (1998). Alliances and networks. *Strategic Management Journal, 19*(4), 293-317.

Gulati, R., Nohria, N., & Zaheer, A. (2000). Strategic networks. *Strategic Management Journal, 21,* 203-215.

Hamel, G. (1991, Winter). Competition for competence and inter-partner learning within international strategic alliances [Special issue]. *Strategic Management Journal, 12,* 83-104.

Hamel, G., Doz, Y., & Prahalad, C. (1989). Collaborate with your competitors and win. *Harvard Business Review, 67,* 133-139.

Hennart, J. F. (1988). A transaction costs theory of equity joint ventures. *Strategic Management Journal, 9*(4), 361-374.

Hoskisson, R. E., Hitt, M. A., Wan, W. P., & Yiu, D. (1999). Theory and research in strategic management: Swings of a pendulum. *Journal of Management, 25*(3), 417-456.

Jarillo, J. C. (1988). On strategic networks. *Strategic Management Journal, 9,* 31-41.

Jarillo, J. C. (1993). *Strategic networks: Creating the borderless organization.* Great Britain: Butterworth-Heinemann.

Kogut, B. (1988). Joint ventures: Theoretical and empirical perspectives. *Strategic Management Journal, 9*(4), 319-332.

Kumar, K., & van Dissel, H. G. (1996, September). Sustainable collaboration: Managing conflict and cooperation in interorganizational systems. *MIS Quarterly,* 279-300.

Limerick, D., & Cunnington, B. (1993). *Managing the new organization: A blueprint for networks and strategic alliances.* San Francisco: Jossey-Bass Publisher.

Osborn, C. S. (1998). Systems for sustainable organizations: Emergent strategies, interactive controls and semi-formal information. *Journal of Management Studies, 35*(4), 481-509.

Osborn, R. N., & Hagedoorn, J. (1997). The institutionalization and evolutionary dynamics of interorganizational alliances and networks. *Academy of Management Journal, 40*(2), 261-278.

Parkhe, A. (1993). Strategic alliance structuring: A game theoretic and transaction cost examination of interfirm cooperation. *Academy of Management Journal, 36,* 794-829.

Porter, M. E., & Millar, V. E. (1986). Cómo obtener ventajas competitivas por medio de la información. *Harvard Deusto Business Review, 25,* 3-20.

Powell, W. W. (1990). Neither market nor hierarchy: Network forms of organization. In B. M. Staw & L. L. Cummings (Eds.), *Research in organizational behavior, 12*, 295-336. Greenwich, CT: JAI Press.

Rackoff, N., Wiseman, C., & Ullrich, W. (1985). IS is for competitive advantage: Implementation of a planning process. *MIS Quarterly, 9,* 285-294.

Ring, P. S., & Van de Ven, A. H. (1994). Developmental processes of cooperative interorganizational relationships. *Academy of Management Review, 19*(1), 90-118.

Smith, K. G., Carroll, S. J., & Ashford, S. J. (1995). Intra- and interorganizational cooperation: Toward a research agenda. *Academy of Management Journal,* (1), 7-23.

Thompson, J. D. (1967). *Organizations in action: Social science bases of administrative theory*. New York: McGraw-Hill.

Thorelli, H. B. (1986). Networks: Between markets and hierarchies. *Strategic Management Journal, 7,* 37-51.

Williamson, O. E. (1975). *Markets and hierarchies: Analysis and antitrust implications*. New York: Free Press.

Williamson, O. E. (1985). *The economic institution of capitalism*. New York: Free Press.

Williamson, O. E. (1991). Comparative economic organization: The analysis of discrete structural alternatives. *Administrative Science Quarterly, 36,* 269-296.

Zenger, T. R., & Hesterly, W. S. (1997). The disaggregation of corporations: Selective intervention, high-powered incentives, and molecular units. *Organization Science, 8*(3), 209-222.

Zenger, T. R., & Marshall, R. (1995). Group-based pay plans: An empirical test of the relationships among size, incentive-intensity, and performance. *Academy of Management Best Paper Proceedings* (pp. 161-165).

Endnote

1 Using different terms, the organization theory literature has examined relationships between firms extensively in various works (some of the most cited are: Contractor & Lorange, 1988; Gulati, 1995; Hamel, 1991; Hamel, Doz, & Prahalad, 1989; Hennart, 1988; Kogut, 1988; Parkhe, 1993; Powell, 1990; Ring & Van de Ven, 1994; Thompson, 1967; Williamson, 1991),, as well as, in some monographic editions of the main academic journals, such as the *Strategic Management Journal* (volume 21, March 2000). We should note the diversity of topics studied in this area, as well as the multi-paradigmatic approach adopted (García Falcón & Medina Muñoz, 1998).

Acknowledgments

The editors would like to acknowledge the help of all those involved in the collation and review process of this book, without whose support the project could not have been satisfactorily completed. A further special note of thanks goes, also, to all the staff at Idea Group Inc., whose contribution throughout the whole process from inception of the initial idea to final publication has been invaluable.

Deep appreciation and gratitude is due to Plan Andaluz de Investigación (project: SEJ-360), for its ongoing sponsorship in terms of financing.

Furthermore, most of the chapter authors also served as referees for articles written by other researchers. Thanks go to all those who provided constructive and comprehensive reviews.

In closing, we wish to thank all of the authors for their insight and excellent contributions to this book. We would also like to thank all of the people who assisted us in the review process. We would particularly like to thank Frank Callan for his invaluable and professional collaboration in the edition of this book. His profound knowledge of English style and grammar has helped us to improve a book that, although in English, was written by researchers from various countries, and aimed at researchers, students and practitioners from around the world. Finally, this book would not have been possible without the ongoing professional support from Mehdi Khosrow-Pour, Renee Davies, Michele Rossi, and Kristin Roth at Idea Group Inc.

Salustiano Martínez-Fierro, PhD

José Aurelio Medina-Garrido, PhD

Jose Ruiz-Navarro, PhD

Cádiz, Spain

March 2006

Chapter I

Alliances and Information Technology: Fundamentals and State of the Art

José Aurelio Medina-Garrido, University of Cádiz, Spain

Salustiano Martínez-Fierro, University of Cádiz, Spain

Jose Ruiz-Navarro, University of Cádiz, Spain

Abstract

This chapter is organized into three sections. In the first section we briefly define some basic concepts about alliances, their attributes, the differences between them and traditional structures, and the types of alliance. The second section analyzes the state of the art in the literature on alliances. We stress the conceptual evolution in organizational design and the main theoretical approaches explaining alliance formation. The final section analyzes the role of information technology (IT) in the formation and running of alliances. In this respect, we stress the impact of IT on organizational structures in general, and on inter-organizational relationships in particular, with these relationships being based on the so-called inter-organizational systems.

Alliance Fundamentals

As early as the 1930s the pioneering research of Roesthlisberger and Dickson (1939) described and stressed the importance of networks of informal relationships within organizations. Subsequently, Miles and Snow (1978) analyzed how market forces give form to structures that diverge from the traditional ones, making organizations more efficient and improving their capacity of reaction. In later publications, market concepts were extended to internal work teams (Miles & Rosenberg, 1982) as well as to external groups, which Miles and Snow (1984) label *dynamic networks*.

These authors signal the 1980s as the point the movement towards the network form began to become apparent, when international competition and rapid technological change forced companies and industries in the United States to embark on drastic restructuring. Firms were then forced to reduce their size (downsizing) to concentrate on their core competencies, flattening (de-layering) the organization by eliminating managers from the hierarchy and externalizing activities. The new firms focused less on growth through vertical integration and instead sought alliances with their suppliers and distributors. Rather than use plans, timetables, and transfer prices to coordinate internal units, they relied on contracts and other exchange agreements connecting autonomous external units in network structures (Miles & Snow, 1992). At this time authors began to describe alliances in more detail (Miles & Snow, 1986; Thorelli, 1986).

From the end of the 1980s onward there was a proliferation of books and articles analyzing the subject of inter-organizational relationships in a great variety of forms. Among these contributions, some of the most important were from Peter Drucker (1990); William Halal (1986); Charles Handy (1990); Russell Johnson and Paul Lawrence (1988); Rosabeth Kanter (1989); and Robert Reich (1991).

We should note the wide diversity of terms that have been used in the literature to refer to the different types of inter-organizational relationship (García Falcón & Medina Muñoz, 1998): inter-organizational ties, organizational interdependence, joint decision making, relational governance, outsourcing, cooperation, coordination, collaboration, collaborative exchange relationship, strategic alliance, joint ventures, inter-firm networks, and strategic networks. In this chapter we have opted to cut down this terminological jungle around the concept of alliance, and more specifically, we concentrate on alliances of a strategic nature, in the terms that we define shortly.

Firms belong to networks of social, professional, and exchange relationships together with other companies (Granovetter, 1985; Gulati, 1998). When these networks are formalized they can take shape in alliances that imply a voluntary agreement between two or more firms involving continuous exchanges; sharing capital, technology, assets, or jointly developing new products or technologies (Ireland, Hitt, Camp, & Sexton, 2001).

According to Jarillo (1993), the concepts of cooperation and competition do not inevitably have to be associated with intra- or inter-organizational behaviors, respectively. Identifying cooperation within the organization and competition in inter-organizational relationships depends on how firms' boundaries are defined. Moreover, these boundaries do not necessarily coincide with the legal boundaries of the firm.

Strategic Alliances

Strategic alliances are stable agreements through which companies establish a network of close relationships that are system ready to provide products and services in a coordinated (Jarillo, 1993) and flexible way (Miles & Snow, 1992). An organization that works in an alliance concentrates on its core capabilities and obtains the complementary resources and capabilities it needs through alliances and outsourcing deals (McGee, 1999; Miles & Snow, 1995). These strategic alliances consist of lasting links that represent an alternative to the hierarchical or market forms of governance (Jarillo, 1988; Thorelli, 1986), and that have strategic significance for the firms entering in the alliances. These links can materialize in strategic alliances; joint ventures; long-term collaboration relationships between suppliers and buyers; and a large number of similar ties (Arndt, 1979; Gulati, Nohria, & Zaheer, 2000). In this type of agreement the transactions are planned and administered rather than managed on an ad hoc basis (Arndt, 1979). However, in spite of the aforementioned planning, companies belonging to alliances coordinate themselves more like markets do than a chain of command would do (Miles & Snow, 1992).

Alliance Attributes

Some of the most important attributes to bear in mind when considering an alliance of companies are: position, power, trust, and critical mass (Thorelli, 1986).

In abstract terms alliances can be seen as a set of nodes and connections between these nodes. A firm's position in the alliance eventually has as much strategic significance as the position of its products in the market. A concept determining the centrality in alliances is power, referring to the ability to influence other alliance members' decisions or actions. An attribute that is closely related to power is *trust*, which assumes that in order to fulfill the commitments undertaken one of the parties in the relationship is ready to do what the other would do if it possessed the former party's resources. Finally, the critical mass refers to the need for an existing relationship—one of particular quantity and quality—if there is going to be an alliance. (Hammarkvist, Håkansson, & Mattsson, 1983).

Differences between Alliances and Traditional Structures

As can be seen in Table 1, these new organizational forms in alliances differ from the traditional forms in many internal and external dimensions. These dimensions transform the traditional organization described by the tree metaphor, with its common communications trunk connecting smaller branches up to the top of the tree, representing the control by the upper management. Instead, alliances can be described by the metaphor of the nervous system, consisting of a unit with many centers, among which, more than one can govern at any one time (Hedlund, 1986).

Types of Alliances

Hinterhuber and Levin (1994) describe four generic types of alliances (Figure 1) in the function of their structural composition: internal, vertical, horizontal, and diagonal.

Internal alliances, to a certain extent, bear some similarity with the traditional concept of strategic business units, although they are a part of the organization they are granted their own autonomy.

Vertical alliances link a particular firm—normally the owner of a technology, reputation, brand, or the basic knowledge required for the activity—with its suppliers and distributors. Among the most well-known and studied cases in the literature we might mention, are those of Toyota and Benetton. The relationship between a franchisor and its franchisees would be included in this category.

Horizontal alliances link similar firms together, and those with similar markets, in order to develop or exploit a particular technology or penetrate a market segment.

Finally, there are a great variety of diagonal alliances, formed between firms attempting to exploit synergies in order to create new and interdisciplinary markets. An example would be the alliances between computing companies involved in the Internet and movie producers.

Table 1. Dimensions of change that lead organizations to form alliances (adapted from Winch, Gyllstrom, Sauer, & Seror-Märklin, 1997, p. 42; Miles & Snow, 1992)

	Traditional firm	Firm in alliance
Identity	Barriers to protect its physical resources	Relationships for prosperity and leverage of its intangibles
Business orientation	On product, and from inside to outside	On customer, and from outside to inside
Decision making	Centralized, sequential, and structured hierarchically	Decentralized, concurrent, and in organizations working in alliance
Associations	Economic and commercial transactions, independence, control, temporality	Interdependence and long-term involvement, achieving common objectives and interests
Interaction	Transactional	Transformational
Integration	Organizations integrated vertically, large, and managed hierarchically	Structures formed from many smaller firms integrated through projects
Outsourcing	Self-sufficient When not, subcontractors keep to contractual obligations	Concentration on core competencies and outsourcing of non-core ones Subcontractors have more proactive role in alliance
Adaptation to customer	Standardized mass products and services	Individualized products and services
Cost imperative	Economies of scale	Economies of scope
Information	Limited access to information	Information overload (need for tools to manage it and for decision making)
Value chain	Activities carried out in company itself or in free market in exchange for, simply, a certain price	Activities largely carried out by different firms with which close relationships are maintained
Processes	Mass production, sub-optimization, automation	Integration of all actors, from suppliers to end consumers; importance of process innovation
Technology	For support and automation; accounts programs	Enabling, to provide help, for decision making, for communication
Resources and capabilities orientation	Oriented to resources Most resources and capabilities required are possessed by firm itself	Oriented to capabilities Resources and capabilities required are distributed among different firms located along value chain
Work groups	Mono-disciplinary department	Multi-disciplinary groups
Transactions	Ad hoc	Planned and managed
Coordination and control	Administrative processes to manage flow of resources	Mechanisms similar to market's, but with cooperation

The taxonomy of alliances proposed by Miles and Snow (1992) is also interesting. For these authors, alliances can be either stable, dynamic, or internal (see Figure 2).

Some alliances involve suppliers, producers, and distributors in a stable relationship with a long-term perspective. In contrast, other alliances are much more dynamic, with components along the value chain that are brought together for a

Figure 1. Types of alliance according to structural composition (adapted from Hinterhuber & Levin, 1994)

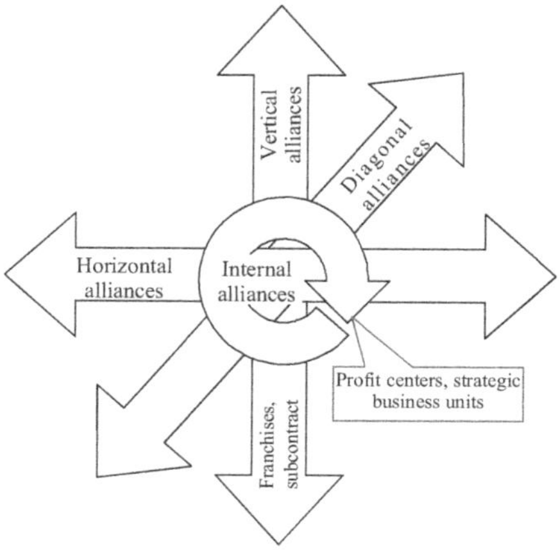

particular project or product, and which will separate once this has ended to join a new value chain for a new business. Some authors label this type of alliance virtual corporation (e.g., Nagel & Dove, 1991).

Finally, within large companies internal alliances appear when managers seek the benefits of market organization by providing their divisions with the possibility of buying and selling both outside and inside the firm.

Lorange and Roos (1992), in turn, argue that alliances can be distinguished in function of: (1) the quantity and the time period of resources committed; and (2) whether the results obtained and the resources committed are kept within the alliance, or alternatively are put at the disposal of the participating companies. By combining these two dimensions, we have the ad hoc pool alliance where firms commit few resources for what is normally a short time period, after which these resources are recovered. When more resources are committed and for a longer time period, but the value created is shared among the participants, the alliance is a consortium. When the resources are few and made available temporarily, but there is an agreement that the value created will be jointly available through a common organization, then this is a project-based strategic alliance. Finally, full-blown strategic alliances are those in which the participants put in abundant resources for a long time period, and the resources generated by the alliance are retained within it.

Figure 2. Miles and Snow's alliance typology (Miles & Snow, 1992)

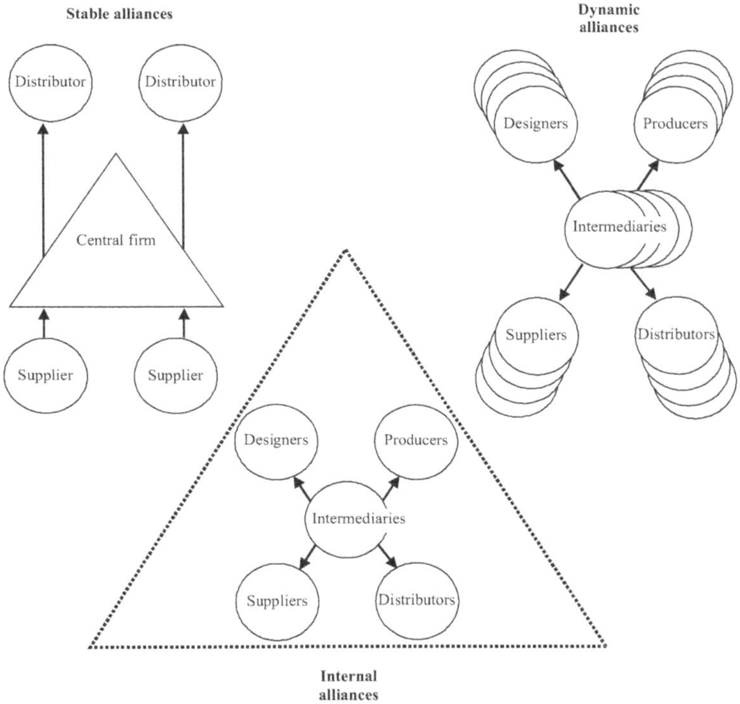

Alliances: State of the Art

In this section we analyze how inter-organizational structures are the result of a historical evolution in ways of organizing companies in accordance with the management knowledge existing at any particular time and the contextual determinants. In this respect, we outline the organizational practices that were the antecedents of alliances.

Subsequently, we study the state of the art in the area of alliances, analyzing the main theories that have been used in the literature. This analysis, conducted from a qualitative perspective, will be completed with a quantitative study in the following chapter.

Evolution in Organizational Design

The evolution in organizational design until it reaches the network form has not been a spontaneous process. It is an evolution that conserves *memories* of the past. Fenton and Pettigrew (2000) organize the evolution of the literature on the design of organizations starting from the universal forms, with which the studies began, and ending with network structures. The process is the result of both continuities and discontinuities in intellectual thinking. There is a common thread connecting functional corporations at the end of the 19th century—divisional and multi-divisional firms and matrix structures (Miles & Snow, 1993; Whittington & Mayer, 1997). The evolution of the literature on organizational design is briefly described in Figure 3.

The first theories on organizational design focused on the development of universal models. In this respect, we would stress the studies on bureaucratic (Weber, 1947) and multi-divisional organizations (Chandler, 1962). The first were based on a mechanistic model of organizational behavior, where the independent variable was the organizational procedures designed to control the activities of the firm's members. The second saw the multi-divisional form as a solution to the problems generated by the growing economies of scale and complexity of industrial companies.

Both the bureaucratic and the multi-divisional organization were universalist, normative models based on the concepts of hierarchy, formalization, division of labor, specialization, and utilitarianism.

In the contingency theories, the Aston studies in England and those by Harvard University in the United States developed the idea of a wide variety of organizational forms in contrast to the earlier universalism. The main interpretation of these studies was that there was no single best way of organizing businesses. Models were developed that sought to identify how the dimensions of the organizational structure related to situational factors such as the environment (Burns & Stalker, 1961; Emery & Trist, 1965; Haige & Aiken, 1967; Lawrence & Lorsch, 1967b; Stinchcombe, 1959), technology (Perrow, 1967, 1970; Woodward, 1965, 1970), or the scale of operations (Blau, 1970; Pugh, Hickson, & Hinings, 1969).

In the organizational design of industrial economics the justification for the multi-divisional form (M-form) was founded on arguments from transaction cost economics (Williamson, 1975). In other words, the M-form separated its strategies and capital into divisions to reduce its transaction costs, creating internal capital markets. This work began with Williamson and Bhargava (1972), and was replicated in the United States (Armour & Teece, 1978; Markides & Williamson, 1996; Teece, 1981), Britain (Ezzamel & Watson, 1993; Grinyer,

Figure 3. Evolution of theoretical perspectives in organizational design (Fenton & Pettigrew, 2000)

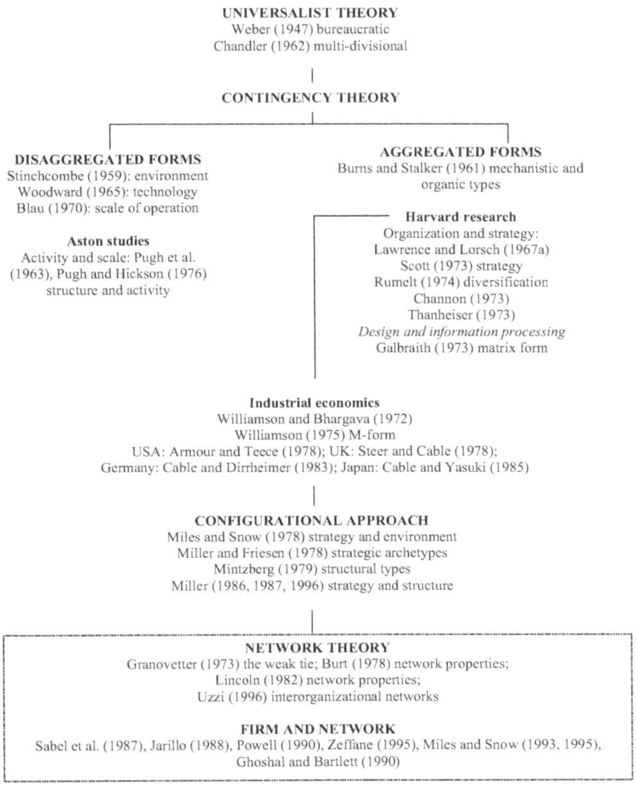

Yasai-Ardekani, & Al-Bazzaz, 1980; Hill, 1988; Hill & Pickering, 1986; Steer & Cable, 1978), Germany (Cable & Dirrheimer, 1983), and Japan (Cable & Yasuki, 1985).

During the 1970s large organizations began to employ complex structural agreements, among which we should mention the matrix form, characterized by a double line of hierarchical responsibility. According to Galbraith (1973), the creation of self-contained tasks reduces the interdependence between units and hence reduces the amount of information that needs to be processed in complex situations.

The decline in interest in the contingent approach was compensated by new theories such as population ecology (Hannan & Freeman, 1977), institutional theory (Powell & DiMaggio, 1991), resource dependency theory (Pfeffer & Salancik, 1978), agency theory (Jensen & Meckling, 1976), and transaction cost economics (Williamson, 1975, 1985).

For its part, the configurational approach also focuses on ideal organizational forms, but attempts to consider the strategy and structure from a more multidimensional perspective. For example, Miles and Snow (1978) show how unique strategic configurations adapt to their environments, and indicate how this could influence the structure and technology of the organization. Miller and Friesen (1977, 1978) also developed a taxonomy of empirical archetypes that link adaptive strategies with their environment and structure. Mintzberg (1979) identifies five structural types, and Miller (1986) reduces these to four structure-strategy combinations.

Population ecology argues that certain common business configurations prevail over others (Aldrich, 1979; Hannan & Freeman, 1977; McKelvey, 1981). This occurs because they are viable, in the sense that they achieve a harmony between their strategy, structure, and context.

The influence of sociology, and in particular, economic sociology, has been predominant in the introduction of network theory concepts within the field of organizational design (Fenton & Pettigrew, 2000). A network approach involves identifying models connecting people and positions on the basis of different types of links (Granovetter, 1992; Lincoln, 1982; Uzzi, 1997).

Networks as a Response to Change

One of the key features of the new context is that the current environment is subjected to a fast pace of change, which some authors have labeled "hyper-competitive" (D'Aveni, 1994; Volberda, 1996), others a "silent industrial revolution" (Prahalad & Hamel, 1994). This new context is characterized by the influence of the new technologies, in particular IT (Halal, 1993; Malone, Morton, & Halperin, 1996; Miles & Snow, 1986), and a faster diffusion of the latest scientific and technological advances (Goldhar & Lei, 1991). Prahalad and Hamel (1994) sum up the forces acting in the competitive environments of the economies of the triad composed by Europe, the United States, and Japan as: deregulation; structural changes due to technological prospects and consumer expectations; excess capacity; growing number of mergers and acquisitions; concern for the environment; reduction in protectionism, technological discontinuities; appearance of trade blocks; and global competition.

The new organization that emerges from this competitive context suffers the breakup of vertically integrated structures and the disintegration of industries (Prahalad & Hamel, 1994); a continuous re-engineering of business processes (Hammer & Champy, 1993); and a focus on the management of these processes rather than on the functional departments (Stewart, 1992). The impact of IT is particularly important, given its role in the management of the information flows both inside and between firms, and because of its influence on the organizational structure (Konstadt, 1990; Navas, 1994).

From this context, a particular set of inter-organizational agreements emerges (Perrone, 1997) in which the central companies share common strategic objectives with a network of organizations (Jarillo, 1993). According to Fenton and Pettigrew (2000), the flexibility of the central firm, combined with the differentiated resources of the network, allows the network-form (N-form) to function in the context of a hyper-competitive environment.

Qualitative Analysis of Theoretical Approaches

In this section we examine the state of the art regarding inter-firm alliances by means of a review of the most significant literature, such as that of Grandori and Soda (1995); Osborn and Hagedoorn (1997); Oliver and Ebers (1998); García Falcón and Medina Muñoz (1998); and Gulati et al. (2000).

According to García Falcón and Medina Muñoz (1998), the literature on inter-firm relationships has been approached from various disciplines, such as marketing, organization theory, organizational economics, and strategic management. To these approaches we would add the sociological perspective. The social perspective, on the functioning of business alliances, has been an important stream in the literature (e.g., Granovetter, 1985; Gulati et al., 2000).

Table 2 sums up the main approaches that have dealt with the topic of alliances, outlining the research questions considered, the most significant authors, and the motives behind the creation of alliances according to them.

We now briefly explain these approaches.

Marketing

The three basic marketing theories applied to the study of inter-organizational relationships are microeconomics, behavior theory, and relationship marketing theory (Stern & Reve, 1980).

Microeconomic theory applies the criterion of economic efficiency to consider what marketing functions could be beneficially carried out by other organizations

(Mallen, 1973). According to this theory, alliances are only the result of the functional interdependencies maintained between two or more organizations.

Behavior theory focuses on the design of mechanisms for controlling the performance of distribution system members (Stern, 1979). Under this approach, inter-firm alliances allow companies to develop and exercise power and coordinate the efforts of participants (Frazier, 1983; Gaski, 1984). The most important topics in inter-organizational relationships have been power (Emerson, 1962; Frazier & Summers, 1984; Stern & Reve, 1980); conflict (Frazier & Summers, 1984) between buyers and sellers (Brown & Day, 1981); communication strategies between sellers and buyers (Mohr & Nevin, 1990); and the determinants of success. These determinants have led to the analysis of commitment, trust, communication, dependence, and inter-organizational conflict as explanatory variables of the success of inter-firm alliances (Ganesan, 1994; Morgan & Hunt, 1994).

The relationship marketing literature studies all the relationships in which the organization participates. Organizations with an adequate marketing orientation will belong to a network of relationships coordinated by norms governing distribution, commitment, and trust (Achrol, 1991). This network of marketing relationships includes both inter- and intra-organizational relationships (Morgan & Hunt, 1994).

Organization Theory

Within organization theory, the theories that have been used most in the analysis of inter-organizational relationships have been contingency theory, resource dependency theory, population ecology, institutional theory (García Falcón & Medina Muñoz, 1998), and the theory of alliance evolution.

Contingency theory examines how firms respond to different variables, particularly environmental ones. Following these premises, there is no single way of organizing the alliances between firms that is universally correct. According to Lorange and Roos (1992), from a contingent perspective firms need to evaluate the resources that will become available if they belong to the network, whether they will be able to recover the resources they themselves make available to the network, and the strategic position that they will occupy with respect to the other members in the network. In turn, Price (1991) maintains that in inter-firm relationships it is important to evaluate the level of standardization of the shared tasks, the legitimate power of each member, the rules governing the relationship, and the relationship between the representatives of the organizations in the network.

Table 2. Theoretical approaches dealing with alliances (adapted from García Falcón & Medina Muñoz, 1998)

Theory	Research topics	Significant authors	Motives behind alliances
Marketing			
Microeconomic theory	Alliances as alternative to internalization and market Analysis of which marketing functions other organizations can undertake	Mallen (1973)	Minimize costs
Behavior theory	Power, conflict, influence, communication, determinants of success	Stern (1979); Frazier (1983); Frazier and Summers (1984); Mohr and Nevin (1990)	Control other organizations Create and exercise power
Relationship marketing theory	Determinants of success Types of marketing relationships	Parvatiyar, Sheth and Whittington (1992); Morgan and Hunt (1994)	Improve coordination
Organization theory			
Contingency theory	Alliances as option of organizational structure Choice of type of alliance Characteristics of alliances: standardization of shared tasks, legitimacy of members, rules of relationship and relationship between representatives of network organizations	Pfeffer (1987); Galbraith (1973); Price (1991); Lorange and Roos (1992)	Maximize efficiency
Resource dependency theory	Alliance as mechanism to increase power and dependence of other organizations in network, and to reduce dependence on environment and uncertainty of exchanges	Pfeffer and Salancik (1978); Oliver (1990)	Maximize power: reduce dependence and uncertainty
Population ecology theory	Survival of network firms compared to isolated firms	Aldrich and Pfeffer (1976); Hannan and Freeman (1977); Aldrich (1979); Astley and Fombrun (1983)	Survive over time in evolving environment
Institutional theory	Isomorphism between allied firms Search for legitimacy belonging to specific networks Social institutions as facilitators of inter-organizational relationships	Pfeffer (1987); Zucker (1987); DiMaggio and Powell (1983); Buchko (1994)	Achieve legitimacy and acceptance
Alliance evolution theory	Strategic alliances as organization strategy; evolving with it, with institutional, organizational, and competitive environment; and with aim of managing alliance itself	Koza and Lewin (1998, 1999); Lewin, Long, and Carroll (1999)	Exploitation of existing capabilities or search for new opportunities

Table 2. continued

Theory	Research topics	Significant authors	Motives behind alliances
Organizational economics			
Transaction cost economics	Network as alternative to market and hierarchy Determinants of decision to participate in network: transaction costs, uncertainty, opportunistic behaviors, bounded rationality, small number of trading partners, information asymmetries, and frequency of transactions	Williamson (1975, 1985, 1991); Pisano, Russo, and Teece (1988); Pisano, Shan, and Teece (1988); Hennart (1988); Buchholtz and Pouder (1991)	Minimize transaction costs
Agency theory	Network as alternative to market and hierarchy Relationships between manufacturers and distributors Inter-organizational coordination and control	Jensen (1983); Anderson (1985); Eisenhardt (1989); Lassar and Kerr (1995)	Minimize agency costs
Strategic management			
Industrial organization	Structural characteristics of alliances (density, structural holes, structural equivalence, and centrality) can affect profitability of firms in an industry Barriers to entry	Porter (1980); Scherer and Ross (1990); Eccles and Crane (1988); Podolny (1993); Baker and Faulkner (1991)	Strategic reasons
Strategic groups	Alliances in strategic blocks Dimension for defining strategic groups Barriers to enter and exit an alliance	Caves and Porter (1977); Cool and Schendel (1988); Nohria and García-Pont (1991)	Strategic reasons
Resource-based view	Obtain, exploit, and develop resources and capabilities through alliances	Mahoney and Pandian (1992); Schulze (1994); Barney (1991); Bergh (1995)	Achieve cooperative and strategic rents
Sociological approaches			
Organizational sociology	Preexisting social relationships as basis for inter-organizational coordination Influence of social norms institutionalized in formation of alliances Organizational culture and its impact on cooperative behavior	Granovetter (1973, 1985); Boisot (1986); Hamilton, Zeile, and Kim (1990); Ring (1993)	Social and cultural embeddedness condition alliance formation
Radical and Marxist studies	Alliances as mechanisms of power	Whitt (1980); Moore (1979); Perucci and Potter (1989)	Reproduce elites and dominant classes
Social network theory	Structure of network: centrality, structural equivalence, and analysis of cliques Dynamic evolution of networks	Benassi (1993); Burt and Minor (1983); Burt (1978, 1979, 1980, 1990, 1993); Gerlach (1992); Laumann, Marsden, and Prensky (1980); Lomi and Grandi (1993); Lomi (1991)	Action of small groups

Resource dependency theory considers that the organization must maximize its power in order to achieve success. In other words, at the same time as it attempts to reduce its dependence on other organizations, it should aim to raise the dependence of these other companies (Pfeffer & Salancik, 1978). Inter-firm alliances are a means of reducing the dependence of alliance members on companies from outside the network as well as raising the power of member companies over external organizations (Cook, 1977; Pfeffer & Salancik, 1978). The starting premise of this theory is that no organization is self-sufficient (Aldrich & Pfeffer, 1976; Pfeffer & Salancik, 1978), and hence, all firms need to build relations with other companies to obtain their resources (Emerson, 1962). Structuring exchanges through alliances with other companies is a way of reducing the uncertainty of exchanges with the environment (Ulrich & Barney, 1984).

Population ecology considers that the environment selects the organizational forms that will prevail (Aldrich, 1979; Hannan & Freeman, 1977), from a longitudinal, rather than cross-sectional perspective (Auster, 1994). Ecological studies on inter-firm alliances have mainly aimed to establish the survival rates of firms in networks compared to isolated companies (Barley, Freeman, & Hybels, 1991; Lomi & Grandi, 1993). They have found some evidence that the network structure in its various forms improves firms' chances of survival.

Institutional theory holds that the behavior of an organization is influenced by its institutional environment. This environment produces a set of norms that dominate practice, and the firm's survival is conditioned by its legitimacy or acceptance in it. For this reason, organizations that share the same environment adopt similar structures and practices over time, a phenomenon known as institutional isomorphism (Meyer & Rowan, 1977; Powell & DiMaggio, 1991). Within this theoretical perspective, scholars analyze how organizations belonging to the same network gradually adopt homogeneous structures and strategies as the network develops (Mezias, 1990; Rowan, 1982). At the same time, firms achieve legitimacy belonging to particular networks (Grandori & Soda, 1995).

Alliance evolution theory takes the alliance itself as its unit of analysis, but it also considers firm-internal elements (like the resource-based view or resource dependency theory), and environmental elements (like institutional theory). Considering alliances as a part of the strategy area and the consequence of the historical evolution of the firm, the theory analyzes their formation and management as a process in which the initial conditions mark the beginning and aim of the relationship, subsequently evolving together with the strategy of the firms' members, and with the organizational, institutional, and competitive environment (Martín & Martínez, 2004).

Organizational Economics

Inter-organizational relationships have been examined from the organizational economics approach mainly through transaction cost economics and agency theory.

In transaction cost economics, Coase (1937) argues that firms exist because the cost of managing economic exchanges between firms (transaction costs) is sometimes higher than the cost of managing the exchanges within companies. Only when contracts are feasible—that is, the parties can be forced to comply with them—and the transaction costs are low, will market exchanges be preferred. The basic premise of transaction cost economics is that markets and hierarchies are alternative governance mechanisms for conducting transactions, and the efficient choice is the one that minimizes the transaction costs (Williamson, 1975, 1985). In this context, Williamson (1991) attempted to incorporate hybrid forms into his conceptual structure as a particular case on the continuum that runs from markets to hierarchies. However, other authors argue that these organizational forms should be regarded as distinct entities from the other two extremes of governance form (Hagedoorn, 1993; Osborn & Hagedoorn, 1997).

Agency theory assumes that human beings are boundedly rational, pursue their own interests, and are opportunistic (Eisenhardt, 1989). Hence, managers (agents) will try to maximize their own utility, even at the expense of the stockholders (principals). By extrapolating the figures of principal and agent to inter-organizational relationships, agency theory can be used to explain how control can be exercised in a network in conditions of diverging objectives between the members and insufficient information (García Falcón & Medina Muñoz, 1998). One of the most discussed topics is the relationship between manufacturers and distributors (Bergen & John, 1991; Lassar & Kerr, 1995; Moorthy, 1987).

Strategic Management

There is growing interest in the field of strategic research regarding the economic impact on firms and their participation in strategic alliances (Gulati et al., 2000).

Among the streams of the strategic management field, the most used in the alliance literature have been the industrial organization school, strategic groups, and the resource-based view.

From the perspective of industrial organization, not only the structural characteristics of the industry (Porter, 1980; Scherer & Ross, 1990), but also the structural characteristics of the alliances can affect the profitability of the companies in an industry. The literature has studied characteristics such as network density, which could lead to oligopolistic coordination (Scherer & Ross, 1990); structural holes, where trading partners of an industry connect through another industry, which can appropriate part of the resources that flow through them (Burt, 1992); and the centrality of the companies in the network, among others. Each of these factors can influence the profitability of the industry and of the firms within it (Gulati et al., 2000).

Strategic groups can be identified on the basis of similarities in the scale of the firm; similarities in their products and services in terms of prices, characteristics and quality; similarities in technologies; or in the customers served. The boundaries of strategic groups have been used to explain the differences in behavior and performance of companies in sectors (Cool & Schendel, 1988). From a network perspective, the attributes of the network can be considered as dimensions for detecting the existence of strategic groups (Gulati et al., 2000). Groups can also be detected within industries (strategic blocks), characterized by the fact that the companies included in these groupings form alliances among themselves, but not with outside companies (Nohria & García-Pont, 1991). It is easy to see how belonging to one block or another produces differences in profitability among the companies of a sector (Piskorski & Nohria, 1999). Another role played by alliances in strategic groups is the creation of barriers to mobility. Thus, the network of relationships of a strategic group from which a firm is attempting to leave, or in which it is attempting to enter, could prevent movements within a sector (Gulati et al., 2000).

The central premise of the resource-based view is that firms achieve competitive advantages due to their heterogeneous bundles of resources (Penrose, 1959). According to Barney (1991), there are four characteristics that a resource must have in order to generate sustainable competitive advantages: value, rareness (scarcity), inimitability, and non-substitutability. Given that inter-firm alliances are idiosyncratic and are created through a path-dependent process (Gulati & Gargiulo, 1999), they are difficult to imitate or substitute. Consequently, so are the idiosyncratic resources and capabilities made available through the unique combination of alliances to which the firm belongs.

From the perspective of resources and capabilities, the search for value in the formation of inter-firm alliances can be differentiated into three generic objectives: (1) to obtain resources and capabilities that are lacking (Guglan & Dunning, 1993; Gulati et al., 2000; Hamel, 1991; Harrigan, 1985; Hill & Hellriegel, 1994; Klein, Crawford, & Alchian, 1978; Kogut, 1988; McGee, 1999; Poppo & Zenger, 1995; Powell, 1987; Prahalad & Hamel, 1990; Shan, Walker, & Kogut, 1994; Teece, 1987); (2) to exploit the resources and capabilities already possessed

(Clemons & Row, 1992; Gulati et al., 2000; Harrigan & Newman, 1990; Kumar & van Dissel, 1996; Tsang, 2000; Vanhonacker, 1997); and (3) to develop resources and capabilities as an option for future strategies (Kogut, 1988).

Sociological Approaches

In parallel with the interest in inter-organizational relationships in the strategy literature, there has been growing interest in understanding how the social context, in which companies are embedded, influences the behavior and performance of these alliances[1] (Gulati et al., 2000).

There are three sociological approaches that have made important contributions to the study of alliances: organizational sociology, radical and Marxist studies, and social network theory.

In the field of organizational sociology there are at least two types of embeddedness that increase the chances that different types of inter-organizational alliances will arise. Using the terms coined by Granovetter (1973, 1985), these two types are social embeddedness and cultural embeddedness. According to Granovetter, every economic relationship between firms takes place within a sphere of preexisting social relationships. This sphere limits the direction as well as the form in which the economic relationship develops. Indeed, some elemental forms of social coordination, such as friendly relations and communication, are the foundation upon which structures emerge for more elaborate inter-organizational coordination (Grandori & Soda, 1995). Institutionalized social norms and values internalized by the economic actors are very likely to affect the appearance of inter-organizational alliances (Hamilton, Zeile, & Kim, 1990). Finally, the organizational culture has been seen as a central phenomenon in the study of the cooperative behavior that emerges between different companies (Ring, 1993).

A stream that is rooted in sociology and in organizational economics is the tradition of radical and Marxist studies. These have looked particularly at those behaviors and strategies of networks that cannot be justified by any concept of efficacy or efficiency, and have instead been explained as pure power mechanisms for reproducing elites and dominant classes (Whitt, 1980). Most of the empirical research on alliances from this perspective has been carried out in the sociology field and has centered on clubs, interlocking directorates, participation in non-commercial institutions, and other informal power networks (Moore, 1979; Perucci & Potter, 1989).

Social network theory has been an important stream of studies on networks within the field of social psychology, fundamentally focusing on the study of small groups. Within the area of inter-organizational relationships, this approach has been applied to the appearance of informal structures and their evolution (Burt, 1990; Lomi, 1991); the boundaries of networks (Burt & Minor, 1983; Laumann et al., 1980); the process of forming committees (cooptation) in the corporation (Burt, 1993); interlocking directorates (Burt, 1979, 1980); and the models of the relationships between small companies (Lomi & Grandi, 1993). The most discussed topic has been the study of the position of individual companies within networks, applying concepts about the structure of the networks such as centrality (Lomi, 1991), structural equivalence (Burt, 1978; Gerlach, 1992; Lomi & Grandi), and the analysis of cliques (Benassi, 1993). Moreover, social network theory has focused on the dynamic evolution of the networks, with the structure of the network at any particular moment in time being predictive of its future evolution (Grandori & Soda, 1995).

Information Technologies and Alliances

IT and Organizational Structure

Advances in IT have been widely recognized as having an impact on organizational structures. However, IT's function has been changing from passively supporting the firm's activities to a more active role in the redesign of organizational structures. In the early days, the rudimentary systems of files and memoranda between offices helped bureaucracies to develop and enable coordination and control between the different components of the organization (Yates, 1989; Yates & Orlikowski, 1992). Later, telephones, the telegraph, and mail systems facilitated distributed organizational forms and inter-organizational communications (Chandler, 1977; de Sola Pool, 1983). Organizational forms were designed to accommodate firms' communications needs. For example, self-contained units were created to reduce environmental uncertainty (Galbraith, 1973; Thompson, 1967; Tushman and Nadler, 1978). On the other hand, IT was designed to adapt to the organizational forms adopted. For example, workflow systems[2] were developed to support the matrix organization (Fulk & DeSanctis, 1995).

Beniger (1986, 1990) maintains that IT and the organization are equivalent forms. IT creates new options for organizational design, and the new organizational forms, in exchange, provide new opportunities for the design of the technology. Hence, a more active vision of IT in the design of the organization is adopted here. Indeed, the design of IT and the design of the organization have become the same task (Lucas & Baroudi, 1994). The articulation of the technologies and the organization makes it clear that neither is preestablished, instead one changes in relation to the other, and the users of the technologies play an active role in this design (Fulk & DeSanctis, 1995).

In 1958, Leavit and Whistler wrote a controversial article predicting that the combination of management science and IT would lead to a reduction in the number of mid-level managers and an increased decentralization of the organization. Since that article was published, a large number of studies have focused on the relation between IT and organizational structure. Some researchers have supported the thesis that IT should be used to centralize the decision-making process and hence reduce the number of mid-level managers (Whistler, 1970). Others have suggested the possibility of increasing decentralization, supporting this argument by noting that IT provides timely and relevant information that can facilitate decentralized decision making and the delegation of power (Arjonilla Domínguez & Medina-Garrido, 2002). It is generally recognized, in any case, that IT is changing the way people work, the structure of organizations, and the structure of entire sectors (Zuboff, 1988; Neumann, 1994).

There is a consensus that IT is extending managers' span of control,[3] reducing the number of hierarchical levels, and making organizations less formalized (García, 1998). According to Drucker (1990) and Daft (1995), IT is flattening organizations, reducing the number of mid-level managers, and making firms more productive and efficient.

IT also helps firms establish either a greater centralization or a greater decentralization, depending on the direction taken by the management, improving coordination even with offices, warehouses, and plants spread out in different parts of the world (García, 1998).

IT also impacts on the organizational structure through the firm's organizational design parameters (Navas, 1994), namely: job design, design of the superstructure, the lateral linkages, and the decision-making system (Mintzberg, 1979).

With regards to job design, IT can be used to reduce the vertical and horizontal specialization[4] of jobs. Workers can thereby develop greater control over their own work, and the work becomes more creative, motivating, and multi-functional. Organizations consequently, become more organic. This occurs because IT is able to take on the routine tasks that traditionally led to a horizontal specialization of jobs, a formalization of tasks, and more bureaucratic structures. This new way of conceiving the division of labor favors a team working over the traditional individualistic concept characteristic of highly specialized jobs.

IT also affects the design of the superstructure, or in other words the design of the authority, which defines the criteria for grouping units. The reduced vertical specialization, combined with team working, cuts the number of managers required, at the same time as the number of hierarchical levels. Another cause of the reduction in these managers is that IT takes on many of the communication, coordination, and control functions for which they were originally responsible as well as widening their span of control. All this leads to flatter organizations.

As far as the unit size is concerned, enriched jobs mean fewer employees are necessary for a particular set of functions, which consequently leads to smaller units (with these understood as groupings of related posts).

According to Mintzberg (1979), the organization uses lateral linkages to coordinate itself. These elements are planning and control, which standardize the outputs and linking mechanisms, which facilitate the mutual adaptation required when direct supervision and standardization are insufficient. IT affects all of these lateral linkages. In the case of planning, it provides greater analytic and design capacity as well as improved access to information, thereby facilitating the capture and subsequent elaboration and handling of the information in formats that are flexible and easy for the decisionmaker to use (Arjonilla Domínguez & Medina-Garrido, 2002). With regards to control, we saw in this section when analyzing the impact of IT on the superstructure that the mid-level managers' need for control is reduced, so that their number diminishes. The new technologies also improve mutual coordination and adaptation, the pillars of the linking mechanisms, by providing more effective communications systems.

Finally, the design of the decision-making system defines the level of centralization or decentralization in the decision making. As we have already mentioned in this section, IT can support both centralization processes (by providing a greater organizational transparency and control capacity to the upper management), and decentralization (or selective decentralization) processes, since information is quickly available from any point in the organization. Hence, IT allows companies to exploit the advantages of centralization and decentralization simultaneously. According to Daft (1995), IT favors both a greater centralization and a greater decentralization, depending on the option chosen by managers.

How Information Technologies Foster Inter-Organizational Relationships

There is broad recognition in the literature that IT can influence the formation of inter-firm alliances (Antonelli, 1992; Bradelley et al., 1993; Child, 1987; Hiltz & Torhoff, 1978; Jarvenpaa & Ives, 1994; Kambil & Short, 1994; Konsynski & McFarlan, 1990; Lucas and Baroudi, 1994; Malone & Rockart, 1991; Nolan et al., 1988).

Researchers attribute IT with the ability to coordinate organizational interdependence within the individual firm. Thus, Rockart and Short (1989) maintain that IT is useful for resolving coordination problems between departments and strategic business units.

Applying these concepts developed for the individual firm, IT has begun to be regarded as a powerful horizontal integrator for managing interdependence between companies (Ciborra, 1990; Ebers, 1993). In this way, it is possible to create inter-organizational systems in which two or more companies share their information systems (Cash & Konsynski, 1985; Porter & Millar, 1986). IT has gained a place among the mechanisms of cooperation between firms, both for the spectacular reduction in communications costs, and for making it possible for geographically dispersed inter-firm alliances to work (Child, 1987; Malone, Yates, & Benjamin, 1987). It is even possible to use computing networks that can, themselves, act as coordination mechanisms—based on machines rather than on people or organizational mechanisms—in an inter-organizational relationship (Grandori & Soda, 1995).

Konsynski and McFarlan (1990) sustain that advances in IT allow companies to cooperate to share data and information, so that the technological and economic risk of large investments can be shared. In this respect, they identify the possibility of creating joint associations in areas such as marketing; intra-sectorial associations; associations between customers and suppliers; and associations promoted by IT suppliers.

Malone et al. (1987), in turn, use transaction cost economics to suggest that IT, by cutting transaction costs, will reduce firms' incentive to integrate vertically and instead induce them to move towards market structures. Clemons and Row (1992) use the same theory to argue that using IT, because it facilitates coordination, generates a network of interacting organizations. Gurbaxani and Whang (1991) observe that IT can either encourage or discourage vertical integration. Ciborra (1993) makes extensive use of transaction cost economics when explaining IT as a facilitator of team, market, and hierarchical structures. Referring to these same two extreme forms of governance—the market and the hierarchy—Zenger and Hesterly (1997) affirm that the recent innovations and advances in performance measurement, control, the design of the organization, and IT have made it possible to emulate market mechanisms within hierarchies and hierarchical mechanisms within markets.

The importance of IT for alliances was already stressed by Thorelli (1986), who identified the planning of management information systems and the services provided by networks of computers as an important strategic implication for the alliance paradigm.

Similarly, Limerick and Cunnington (1993) also argue for the importance of IT, regarding it as the driving force for the expansion of organizational and strategic alliances. Along the same line, Johnson and Lawrence (1988) sustain that current computer processing and communications costs make it possible to predict the competitive advantages of association for small companies, which can each develop one part of the value chain and coordinate their activities with the other alliance members.

We can therefore distinguish a separate concept within the range of information technologies. IT that crosses the boundaries of different organizations can be termed inter-organizational systems. Cash and Konsynski (1985) define these systems as automated information systems that are shared between two or more companies.

Buyers and sellers can use inter-organizational systems to agree on routines for their trade exchanges, doing away with the need for direct negotiations in each transaction (Senn, 1996). These systems allow strategic alliance members to interconnect their businesses. In this way, they can rationalize their business processes in spite of geographic distance, reducing their transaction costs, eliminating paper, and the inefficiencies associated with paperwork (Senn, 1996; Venkatraman & Bensaou, 1996).

Among inter-organizational systems there are a wide variety of technologies, of which the most important are EDI (Electronic Data Interchange), EFT (Electronic Funds Transfer), electronic forms, shared databases, e-mail, computer-aided conferencing, audioconferencing, and videoconferencing.

References

Achrol, R. S. (1991). Evolution of the marketing organization: New forms for turbulent environments. *Journal of Marketing, 55*(4), 77-93.

Aldrich, H. (1979). *Organization and environment.* Englewood Cliffs, NJ: Prentice Hall.

Aldrich, H. E., & Pfeffer, J. (1976). Environment of organizations. *Annual Review of Sociology, 2,* 79-105.

Anderson, E. (1985). The salesperson as outside agent of employee: A transaction cost analysis. *Marketing Science, 4,* 234-245.

Antonelli, C. (1992). *The economics of information networks.* North Holland: Elsevier Science Publisher.

Arjonilla Domínguez, S. J., & Medina-Garrido, J. A. (2002). *La gestión de los sistemas de información en la empresa.* Madrid: Ediciones Pirámide. (*The management of the information system in the company.* Madrid, Spain: Piramide Publishing.)

Armour, H. O., & Teece, D. J. (1978). Organizational structure and economic performance: A test of the multi-divisional hypothesis. *The Ranch Journal of Economics, 9*(1), 106-122.

Arndt, J. (1979). Towards a concept of domesticated markets. *Journal of Marketing, 43*(4), 69-75.

Astley, W. G., & Fombrun, C. J. (1983). Collective strategy: Social ecology of organizational environments. *Academy of Management Review, 8*(4), 576-587.

Auster, E. R. (1994). Macro and strategic perspectives on interorganizational linkages: A comparative analysis and review with suggestions for reorientation. *Advances in Strategic Management, 10B,* 3-40.

Baker, W. E., & Faulker, R. R. (1991). Role as resource in the Hollywood film industry. *American Journal of Sociology, 97,* 279-309.

Barley, S. R., Freeman, J., & Hybels, R. C. (1991). *Strategic alliances in commercial biotechnology.* Working Paper. E.G.O.S. Colloquium, Vienna, July.

Barney, J. B. (1991). Firm resources and sustained competitive advantage. *Journal of Management, 17*(1), 99-120.

Baum, J. A. C., & Dutton, J. (1996). The embeddedness of strategy. In P. Shrivastava, A. S. Huff, & J. E. Dutton (Eds.), *Advances in Strategic Management* (Vol. 13, pp. 3-40). Greenwich, CT: JAI Press.

Benassi, M. (1993). *Dalla gerarchia alla rete: Modelli ed esperienze organizzative.* Milan, Italy: Fondazione IBM. Etas. (*From hierarchy to the network: Models and organizational experiences.* Milan, Italy: IBM Foundation. Etas.)

Beniger, J. R. (1986). *The control revolution: Technological and economic origins of the information society.* Cambridge, MA: Harvard University Press.

Beniger, J. R. (1990). Conceptualizing information technology as organization, and vice versa. In J. Fulk & C. Steinfield (Eds.), *Organizations and communication technology.* Newbury Park, CA: Sage Publications.

Bergen, M., & John, G. (1991). *Understanding the participation rate decision in cooperative advertising plans.* Marketing Technical Report and Reprint Series 87. Chicago, IL: University of Chicago.

Bergh, D. D. (1995). Size and relatedness of units sold: An agency theory and resource-based perspective. *Strategic Management Journal, 16,* 221-239.

Blau, R. M. (1970). A formal theory of differentiation in organizations. *American Sociological Review, 35*(2), 201-218.

Boisot, M. H. (1986). Markets and hierarchies in a cultural perspective. *Organization Studies, 7*(2), 135-158.

Bradelley, S. P., Hausman, J. A., & Nolan, R. L. (1993). *Globalization, technology and competition: The fusion of computers and telecommunications in the 1990s.* Boston: Harvard Business School Press.

Brown, J., & Day, R. (1981). Measures of manifest conflict in distribution channels. *Journal of Marketing Research, 18,* 263-274.

Buchholtz, A., & Pouder, R. (1991). *From hierarchies to markets: A transaction cost analysis of the decision to privatize public services.* Working Paper. Storrs, CT: University of Connecticut.

Buchko, A. A. (1994). Barriers to strategic transformation: Interorganizational networks and institutional forces. *Advances in Strategic Management, 10B,* 81-106.

Burns, T., & Stalker, G. M. (1961). *The management of innovation.* London: Tavistock.

Burt, R. S. (1978). Cohesion versus structural equivalence as a basis for network subgroups. *Sociological Methods and Research, 7,* 189-212.

Burt, R. S. (1979). A structural theory of interlocking corporate directorates. *Social Networks, 1,* 415-435.

Burt, R. S. (1980). Cooptive corporate actor networks: A reconsideration of interlocking directorates involving American manufacturing. *Administrative Science Quarterly, 25,* 557-582.

Burt, R. S. (1990). *A study of structural holes as social capital and entrepreneurial opportunity.* Working Paper. N.I.A.S. Symposium on Interdisciplinary Perspective on Organization Studies.

Burt, R. S. (1992). *Structural holes: The social structure of competition.* Cambridge, MA: Harvard University Press.

Burt, R. S. (1993). *Corporate profits and cooptation.* New York: Academic Press.

Burt, R. S., & Minor, M. (Eds.). (1983). *Applied network analysis.* Beverly Hills, CA: Sage Publications.

Cable, J., & Dirrheimer, M. J. (1983). Hierarchies and markets: An empirical test of the multidivisional hypothesis in West Germany. *International Journal of Industrial Organization, 1,* 43-62.

Cable, J., & Yasuki, H. (1985). Internal organization, business groups and corporate performance: An empirical test of the multi-divisional hypothesis in Japan. *International Journal of Industrial Organization, 3*(4), 401-20.

Cash, J. I., & Konsynski, B. R. (1985). IS redraws competitive boundaries. *Harvard Business Review, 63*(2), 134-142.

Caves, R., & Porter, M. E. (1977, May). From entry barriers to mobility barriers: Conjectural decisions and contrived deterrence to new competition. *Quarterly Journal of Economics, 91*(2), 241-261.

Chandler, A. D. (1962). *Strategy and structure: Chapters in the history of the American industrial enterprise.* Cambridge, MA: MIT Press.

Chandler, A. D. (1977). *The visible hand: The managerial revolution in American business.* Cambridge, MA: Harvard University Press.

Channon, D. (1973). *The strategy and structure of British enterprise.* Cambridge, MA: Harvard University Press.

Child, J. (1987, Fall). Information technology, organization and the response to strategic challenges. *California Management Review, 30*(1), 33-49.

Ciborra, C. (1990). *Tecnologie di coordinamento.* Milan: Angeli. (*Coordination technologies.* Milan, Italy: Angeli.)

Ciborra, C. (1993). *Teams, markets, and systems: Business innovation and information technology.* Cambridge, UK: Cambridge University Press.

Clemons, E. K., & Row, M. C. (1992). Information technology and industrial cooperation: The changing economics of coordination and ownership. *Journal of Management Information Systems, 9*(2), 9-28.

Coase, R. H. (1937). The nature of the firm. *Economica, 4,* 386-406.

Cook, K. (1977). Exchange power in networks of interorganizational relations. *Sociological Quarterly, 18,* 62-82.

Cool, K. O., & Schendel, D. E. (1988). Performance differences among strategic group members. *Strategic Management Journal, 9*(3), 207-223.

Daft, R. L. (1995). *Organizational theory and design.* St. Paul, MN: West Publishing.

D'Aveni, R. (1994). *Hypercompetition: Managing the dynamics of strategic maneuvering.* New York: Free Press.

de Sola Pool, I. (1983). *Forecasting the telephone: A retrospective assessment.* Norwood, NJ: Ablex.

Dimaggio, P. J., & Powell, W.W. (1983). The iron cage revisited: Institutional isomorphism and collective rationality in organizational fields. *American Sociological Review, 48,* 147-160.

Drucker, P. F. (1990). *The new reality*. New York: Harper & Row.

Ebers, M. (1993, July 6-8). *IT networks as organizational forms*. Paper presented at the 11th E.G.O.S. Colloquium, The Production and Diffusion of Managerial knowledge, Paris.

Eccles, R. G., & Crane, D. B. (1988). *Doing deals*. Boston: Harvard Business School Press.

Eisenhardt, K. M. (1989). Agency theory: An assessment and review. *Academy of Management Review, 14,* 57-74.

Emerson, R. M. (1962). Power-dependence relations. *American Sociological Review, 27,* 31-41.

Emery, F. E., & Trist, E. L. (1965). The causal texture of organizational environments. *Human Relations, 18,* 21-32.

Ezzamel, M., & Watson, R. (1993). Organizational form, ownership structure and corporate performance: A contextual empirical analysis of UK companies. *British Journal of Management, 4*(3), 161-176.

Fenton, E. M., & Pettigrew, A. M. (2000). Theoretical perspectives on new forms of organizing. In A. M. Pettigrew & E. M. Fenton (Eds.), *The innovation organization*. London: Sage Publications.

Frazier, G. L. (1983). Interorganizational exchange behavior in marketing channels: A broadened perspective. *Journal of Marketing, 47,* 68-78.

Frazier, G. L., & Summers, J. (1984). Interfirm influence strategies and their application within distribution channels. *Journal of Marketing, 48,* 43-55.

Fulk, J., & DeSanctis, G. (1995). Electronic communication and changing organizational forms. *Organization Science, 6*(4), 337-349.

Galbraith, J. (1973). *Designing complex organizations*. Reading, MA: Addison-Wesley.

Ganesan, S. (1994). Determinants of long-term orientation in buyer-seller relationships. *Journal of Marketing, 58,* 1-19.

García Calderón Díaz, L. (1998). *A PLS model of strategic networks and organizational performance: Based on environmental turbulence, information technology, and transaction costs*. Doctoral thesis, Instituto Tecnológico y de Estudios Superiores de Monterrey, Monterrey, Mexico.

García Falcón, J. M., & Medina Muñoz, D. R. (1998). Enfoques teóricos sobre las relaciones interorganizativas: Una revisión comparativa. (Approaches to interorganizational relationship: A comparative revision.) *Revista Europea de Dirección y Economía de la Empresa, 7*(3), 9-30.

Gaski, J. F. (1984). The theory of power and conflict in channels of distribution. *Journal of Marketing, 48,* 9-29.

Gerlach, M. L. (1992). The Japanese corporate network: A blockmodel analysis. *Administrative Science Quarterly, 37,* 105-139.

Ghoshal, S., Bartlett, C. A. (1990). The multinational corporation as an interorganizational network. *Academy of Management Review, 15*(4), 603-625.

Goldhar, J. D., & Lei, D. (1991). The shape of twenty-first century global manufacturing. *Journal of Business Strategy, 12*(2), 37-41.

Grandori, A., & Soda, G. (1995). Inter-firm network: Antecedents, mechanisms and forms. *Organization Studies, 16*(2), 183-214.

Granovetter, M. S. (1973). The strength of weak ties. *American Journal of Sociology, 78*(6), 1360-1380.

Granovetter, M. S. (1985). Economic action and social structure: The problem of embeddedness. *American Journal of Sociology, 91,* 481-510.

Granovetter, M. S. (1992). Problems of explanation in economic sociology. In N. Nohria & R. G. Eccles (Eds.), *Networks and organizations: Structure, form and action* (pp. 25-26). Boston: Harvard Business School Press.

Grinyer, P. H., Yasai-Ardekani, M., & Al-Bazzaz, S. (1980). Strategy, structure, the environment and financial performance in 48 United Kingdom companies. *Academy of Management Journal, 23*(2), 193-220.

Guglan, P., & Dunning, J. H. (1993). Technology based cross-border alliances. In R. Culpan (Ed.), *Multinational strategic alliances.* Binghamton, NY: Howarth Press Inc.

Gulati, R. (1998). Alliances and networks. *Strategic Management Journal, 19*(4), 293-317.

Gulati, R., & Gargiulo, M. (1999). Where do interorganizational networks come from? *American Journal of Sociology, 104*(5), 1439-1493.

Gulati, R., Nohria, N., & Zaheer, A. (2000). Strategic networks. *Strategic Management Journal, 21,* 203-215.

Gurbaxani, V., & Whang, S. (1991). The impact of information systems on organizations and markets. *Communications of the ACM, 34*(1), 59-73.

Hagedoorn, J. (1993). Understanding the rationale of strategic technology partnering: International modes of cooperation and sectoral differences. *Strategic Management Journal, 14,* 371-385.

Haige, J., & Aiken, M. (1967). Relationship of centralization to other structural properties. *Administrative Science Quarterly, 12,* 72-92.

Halal, W. (1986). *The new capitalism.* New York: John Wiley & Sons.

Halal, W. (1993). The transition from hierarchy to... what? In W. Halal, A. Geranmayeh, & J. Pourdehnad (Eds), *Internal markets: Bringing the power of free enterprise inside your organization* (pp. 27-51). New York: John Wiley & Sons.

Hamel, G. (1991, Winter). Competition for competence and inter-partner learning within international strategic alliances [Special issue]. *Strategic Management Journal, 12,* 83-104.

Hamilton, G., Zeile, W., & Kim, W. J. (1990). The network structure of East Asian economies. In S. R. Clegg & G. Redding (Eds.), *Capitalism in contrasting cultures* (pp. 105-129). Berlin: De Gruyter.

Hammarkvist, K.-O., Håkansson, H., & Mattsson, L.-G. (1983). *Marknadsföring för konkurrenskraft.* [Marketing for competitiveness]. Malmo, Sweden: Liber.

Hammer, M., & Champy, J. (1993). *Reengineering the corporation.* New York: Harper Business.

Handy, C. (1990). *The age of unreason.* Cambridge, MA: Harvard Business School Press.

Hannan, M. T., & Freeman, J. (1977). The population ecology of organizations. *American Journal of Sociology, 83,* 929-964.

Harrigan, K. R. (1985). *Strategy flexibility.* Lexington, MA: Lexington Books.

Harrigan, K. R., & Newman, W. H. (1990). Bases of interorganization co-operation: Propensity, power, persistence. *Journal of Management Studies, 27*(4), 417-434.

Hedlund, G. (1986). The hypermodern MNC—A heterarchy? *Human Resource Management, 25*(1), 9-35.

Hennart, J. F. (1988). A transaction costs theory of equity joint ventures. *Strategic Management Journal, 9*(4), 361-374.

Hill, C. W. L. (1988). Internal capital markets controls and financial performance in multidivisional firms. *Journal of Industrial Economics, 37,* 67-83.

Hill, C. W. L., & Pickering, J. F. (1986). Divisionalisation, decentralisation and performance in large United Kingdom companies. *Journal of Management Studies, 23,* 26-50.

Hill, R.C., & Hellriegel, D. (1994). Critical contingencies in joint ventures management: Some lessons from managers. *Organization Science, 5,* 594-607.

Hiltz, R. S., & Torhoff, M. (1978). *Network nation: Human communications via computers.* Reading, MA: Addison-Wesley.

Hinterhuber, H. H., & Levin, B. M. (1994). Strategic networks—The organization of the future. *Long Range Planning, 27*(3), 43-53.

Ireland, R. D., Hitt, M. A., Camp, S. M., & Sexton, D. L. (2001). Integrating entrepreneurship and strategic management actions to create firm wealth. *Academy of Management Executive, 15*(1), 49-63.

Jarillo, J. C. (1988). On strategic networks. *Strategic Management Journal, 9,* 31-41.

Jarillo, J. C. (1993). *Strategic networks: Creating the borderless organization.* UK: Butterworth-Heinemann.

Jarvenpaa, S. L., & Ives, B. (1994). The global network organization of the future: Information management opportunities and challenges. *Journal of Management Information Systems, 10*(4), 25-57.

Jensen, M. C. (1983). Organization theory and methodology. *Accounting Review, 56,* 319-338.

Jensen, M. C., & Meckling, W. H. (1976). Theory of the firm: Managerial behaviour, agency costs and ownership structure. *Journal of Financial Economics, 3,* 305-360.

Johnson, R., & Lawrence, P. R. (1988, July-August). Beyond vertical integration: The rise of the value-adding partnership. *Harvard Business Review,* 94-101.

Kambil, A., & Short, J. E. (1994). Electronic integration and business network redesign: A roles-linkage perspective. *Journal of Management Information Systems, 10*(4), 59-83.

Kanter, R. M. (1989). *When giants learn to dance: Mastering the challenges of strategy, management and careers in the 1990's.* New York: Simon & Schuster.

Klein, B., Crawford, R. G., & Alchian, A. A. (1978). Vertical integration, appropriable rents, and the competitive contracting process. *Journal of Law and Economics, 21*(2), 297-326.

Kogut, B. (1988). Joint ventures: Theoretical and empirical perspectives. *Strategic Management Journal, 9*(4), 319-332.

Konstadt, P. (1990, August). Into the breach. *CIO, 11*, 71-73.

Konsynski, M. (1990). Information partnerships—Shared data, shared scale. *Harvard Business Review, 68*(5), 114-120.

Koza, M. P., & Lewin, A. Y. (1998). The coevolution of strategic alliances. *Organizational Science, 9*(3), 255-264.

Koza, M. P., & Lewin, A. Y. (1999). The coevolution of network alliances: A longitudinal analysis of an international professional service network. *Organizational Science, 10*(5), 638-653.

Kumar, K., & van Dissel, H. G. (1996, September). Sustainable collaboration: Managing conflict and cooperation in interorganizational systems. *MIS Quarterly*, 279-300.

Lassar, W. M., & Kerr, J. L. (1995). *Strategy and control in supplier-distributor relationships: An agency perspective.* Working Paper. Coral Gables, FL: University of Miami.

Laumann, E., Marsden, P., & Prensky, D. (1980). The boundary specification problem in network analysis. In R. Burt & M. Minor (Eds.), *Applied network analysis* (pp. 18-34). Beverly Hills, CA: Sage Publications.

Lawrence, P. R., & Lorsch, J. W. (1967a). Differentiation and integration in complex organizations. *Administrative Science Quarterly, 12*, 1-47.

Lawrence, P. R., & Lorsch, J. W. (1967b). *Organization and environment: Managing differentiation and integration.* Boston: Harvard University, Division of Research, Graduate School of Business.

Leavit, H. J., & Whistler, T. L. (1958). Management in the 1980's. *Harvard Business Review, 36*(6), 41-48.

Lewin, A. Y., Long, C. P., & Carroll, T. N. (1999). The coevolution of new organizational forms. *Organizational Science, 10*(5), 535-550.

Limerick, D., & Cunnington, B. (1993). *Managing the new organization: A blueprint for networks and strategic alliances.* San Francisco: Jossey-Bass Publisher.

Lincoln, J. R. (1982). Intra-(and inter-) organizational networks. In S. B. Bacharach (Ed.), *Research in the sociology of organizations* (Vol. 1, pp. 1-38). Greenwich, CT: JAI Press.

Lomi, A. (1991). *Reti organizzative.* Bologna: Il Mulino. (*Organizational networks*. Bologna: Il Mulino.)

Lomi, A., & Grandi, A. (1993, September). *The network structure of interorganizational communities: Preliminary evidence from a study of inter-firm relations in the southern Italian mechanical industry.* Paper presented at the European Science Foundation Conference Forms of Interorganizational Networks: Structures and Processes, Berlin.

Lorange, P., & Roos, J. (1992). *Strategic alliances: Formation, implementation and evolution.* Oxford: Blackwell.

Lucas, Jr., H. C., & Baroudi, J. (1994). The role of information technology in organization design. *Journal of Management Information Systems, 10*(4), 9-23.

Mahoney, J., & Pandian, R. (1992). The resource-based view within the conversation of strategic management. *Strategic Management Journal, 13,* 363-380.

Mallen, B. (1973). Functional spin-off: A key to anticipating changes in distribution structure. *Journal of Marketing, 13,* 18-25.

Malone, T. W., Morton, M. S. S., & Halperin, R. R. (1996). Organizing for the 21st century. *Strategy and Leadership, 24*(4), 6-10.

Malone, T. W., & Rockart, J. F. (1991). Computers, networks, and the corporation. *Scientific American, 265*(3), 128-137.

Malone, T. W., Yates, J., & Benjamin, R. I. (1987). Electronic markets and electronic hierarchies. *Communications of the ACM, 30*(6), 484-497.

Markides, C. C., & Williamson, P. J. (1996). Corporate diversification and organizational structure: A resource-based view. *Academy of Management Journal, 39*(2), 340-368.

Martín-Alcázar, F., Martínez-Fierro, S. (2004). El proceso de alianzas estratégicas desde el enfoque de la teoría evolutiva de alianzas. (The process of strategic alliances from the approach of the evolutionary theory of alliances.) *Investigaciones Europeas de Dirección y Economía de la Empresa, 10*(3), 1135-2523.

McGee, J. (1999). *Knowledge dynamics and corporate restructuring: Deconstructionism as a new paradigm?* Paper presented at the 2nd International Workshop on Organization of the Future in the "Information Society": Managing Change, Human Resources and Structure, Cadiz, Spain.

McKelvey, B. (1981). *Organizational systematics.* Los Angeles: University of California Press.

Meyer, J. W., & Rowan, B. (1977). Institucionalized organizations: Formal structure as myth and ceremony. *American Journal of Sociology, 83,* 340-363.

Mezias, S. J. (1990). An institutional model of organization practice: Financial reporting at the fortune 200. *Administrative Science Quarterly, 35,* 431-457.

Miles, R. E., & Rosenberg, H. (1982). The human resources approach to management. *Organizational Dynamics, 10*(3), 26-41.

Miles, R. E ., & Snow, C. C. (1978). *Organizational strategy, structure and process*. New York: McGraw-Hill.

Miles, R. E., & Snow, C. C. (1984). Fit, failure and the hall of fame. *California Management Review, 26*(3), 10-28.

Miles, R. E., & Snow, C. C. (1986). Network organizations: New concepts for new forms. *California Management Review, 28,* 62-73.

Miles, R. E., & Snow, C. C. (1992, Summer). Causes of failure in network organizations. *California Management Review, 34*(4), 53-72.

Miles, R. E., & Snow, C. C. (1993). Internal markets and network organizations. In W. E. Halal, A. Geranmayeh, & J. Pardehnad (Eds.), *Internal markets: Bringing the power of free enterprise inside your organization* (pp. 67-86). New York: John Wiley.

Miles, R. E., & Snow, C. C. (1995). The new network firm: A spherical structure built on a human investment philosophy. *Organizational Dynamics, 23*(4), 5-17.

Miller, D. (1986). Configurations of strategy and structure: Towards a synthesis. *Strategic Management Journal, 7,* 233-249.

Miller, D. (1987). The genesis of configuration. *Academy of Management Review, 12*(4), 686-701.

Miller, D. (1996). Configurations revisited. *Strategic Management Journal, 17,* 505-512.

Miller, D., & Friesen, P. H. (1977). Strategy making in context: Ten empirical archetypes. *Journal of Management Studies, 14,* 259-280.

Miller, D., & Friesen, P. H. (1978). Archetypes of strategy formulation. *Management Science, 24,* 921-933.

Mintzberg, H. (1979). *The structuring of organizations*. Englewood Cliffs, NJ: Prentice Hall. Translated in 1995: *La estructuración de las organizaciones*. Ariel economía, Barcelona, Spain.

Mohr, J., & Nevin, R. (1990). Communications strategies in marketing channels: A theoretical perspective. *Journal of Marketing, 54,* 36-51.

Moore, G. (1979). The structure of a national elite network. *American Sociological Review, 44,* 673-692.

Moorthy, K. S. (1987). Managing channels profits: Comment. *Marketing Science, 6,* 375-379.

Morgan, R. M., & Hunt, S. D. (1994). The commitment-trust theory of relationship marketing. *Journal of Marketing, 58,* 20-38.

Nagel, R., & Dove, R. (1991). *21st century manufacturing enterprise strategy*. Lehigh: Iacocca Institute of Lehigh University.

Navas López, J. E. (1994). *Organización de la empresa y nuevas tecnologías*. Madrid: Pirámide. (*Company organization and new technologies*. Madrid, Spain: Pirámide.)

Neumann, S. (1994). *Strategic information systems: Competition through information technologies*. New York: Macmillan College Publication Company.

Nohria, N., & García-Pont, C. (1991, Summer). Global strategic linkages and industry structure [Special issue]. *Strategic Management Journal, 12,* 105-124.

Nolan, R. L., Pollock, A. J., & Ware, J. P. (1988). Creating the 21st century organization. *Stage-by-Stage, 8*(4), 1-11.

Oliver, A. L., & Ebers, M. (1998). Networking network studies: An analysis of conceptual configurations in the study of inter-organizational relationships. *Organization Studies, 19*(4), 549-583.

Oliver, C. (1990). Determinants of interorganizational relationships: Integration and future directions. *Academy of Management Review, 15,* 241-265.

Osborn, R. N., & Hagedoorn, J. (1997). The institutionalization and evolutionary dynamics of interorganizational alliances and networks. *Academy of Management Journal, 40*(2), 261-278.

Parvatiyar, A., Sheth, J. N., & Whittington, Jr., B. (1992). *Paradigm shift in interfirm marketing relationships: Emerging research issues*. Working Paper. Atlanta, GA: Emory University.

Penrose, E. T. (1959). *The theory of the grow of the firm*. New York: John Wiley & Sons.

Perrone, V. (1997). The coevolution of contexts and structures: The N-form. In T. Clark (Ed.), *Organizational behaviour: Essays in honour of Derek S. Pugh* (pp. 145-163). Aldershot; Hants, UK: Ashgate.

Perrow, C. (1967). A framework for the comparative analysis of organizations. *American Sociological Review, 32,* 194-208.

Perrow, C. (1970). *Organizational analysis: A sociological view*. London: Tavistock.

Perucci, R., & Potter, H. R. (1989). *Networks of power*. Berlin: De Gruyter.

Pfeffer, J. (1987). *Organizaciones y teoría de la organización.* Buenos Aires: El Ateneo Editorial. (*Organizations and theory of the organization.* Buenos Aires: Ateneo Publishing.)

Pfeffer, J., & Salancik, G. R. (1978). *The external control of organizations: A resource dependence perspective.* New York: Harper & Row.

Pisano, G., Russo, M., & Teece, D. (1988). Joint ventures and collaborative arrangements in the telecommunications equipment industry. In D. Mowery (Ed.), *International collaborative ventures in U.S. manufacturing.* Cambridge, MA: Ballinger Press.

Pisano, G., Shan, W., & Teece, D. (1988). Joint ventures and collaboration in the biotechnology industry. In D. Mowery (Ed.), *International collaborative ventures in U.S. manufacturing.* Cambridge, MA: Ballinger Press.

Piskorski, M., & Nohria, N. (1999). *Allocation to open and closed portfolios.* Working Paper. Boston: Harvard Business School.

Podolny, J. (1993). A status based model of market competition. *American Journal of Sociology, 98,* 829-872.

Poppo, L., & Zenger, T. (1995). Opportunism, routines, and boundary choices: A comparative test of transaction cost and resource-based explanations for make-or-buy decisions [Best papers proceedings]. *Academy of Management Journal,* 42-46.

Porter, M. E. (1980). *Competitive strategy.* New York: Free Press.

Porter, M. E., & Millar, V. E. (1986). Cómo obtener ventajas competitivas por medio de la información. (How to obtain competitive advantages by means of the information.) *Harvard Deusto Business Review, 25,* 3-20.

Powell, W. W. (1987). Hybrid organizational arrangements: New forms or transitional development? *California Management Review, 30*(1), 67-87.

Powell, W. W. (1990). Neither market nor hierarchy: Network forms of organization. In B. M. Staw & L. L. Cummings (Eds.), *Research in organizational behavior* (Vol. 12, 295-336). Greenwich, CT: JAI Press.

Powell, W. W., & Dimaggio, P. J. (1991). *The new institutionalism in organizational analysis.* Chicago: The University of Chicago Press.

Prahalad, C. K., & Hamel, G. (1990). The core competence of the corporation. *Harvard Business Review, 68*(3), 79-91.

Prahalad, C. K., & Hamel, G. (1994). Strategy as a field of study: Why search for a new paradigm? *Strategic Management Journal, 15,* 5-16.

Price, R. A. (1991). An investigation of path-goal leadership theory in marketing channels. *Journal of Retailing, 67*(3), 339-361.

Pugh, D. S., & Hickson, D. J. (1976). *Organizational structure in its context: The Aston programme 1.* Farnborough, Hants: Saxon House; Lexington, MA: Lexington Books.

Pugh, D. S., Hickson, D. J., Hinings, C. R., Macdonald, K., Turner, C., & Lupton, T. (1963). A conceptual scheme for organizational analysis. *Administrative Science Quarterly, 8*, 289-315.

Pugh, D. S., Hickson, D. J., & Hinings, C. R. (1969). The context of organization structures. *Administrative Science Quarterly, 14*, 47-61.

Reich, R. (1991). *The work of nations: Preparing ourselves for 21st century capitalism.* New York: A.A. Knopf.

Ring, P. (1993). *Processes facilitating reliance on trust in inter-organizational networks.* Paper presented at the European Science Foundation Conference, Forms of Interorganizational Networks: Structures and Processes. Berlin.

Rockart, J. F., & Short, J. E. (1989, Winter). IT in the 1990s: Managing organizational interdependence. *Sloan Management Review,* 7-17.

Roesthlisberger, F. F., & Dickson, W. J. (1939). *Management and the worker.* Cambridge, MA: Harvard University Press.

Rowan, B. (1982). Organizational structure and the institutional environment: The case of public schools. *Administrative Science Quarterly, 7,* 259-279.

Rumelt, R. (1974). *Strategy, structure and economic performance.* Boston: Harvard Business School Press.

Sabel, C., Herrigel, G., Kazis, R., & Deeg, R. (1987). How to keep mature industries innovative. *Technology Review, 90*(3), 27-35.

Scherer, F. M., & Ross, D. (1990). *Industrial market structure and economic performance.* Boston: Houghton Mifflin.

Schulze, W. E. (1994). The two schools of thought in resource-based theory: Definitions and implications for research. *Advances in Strategic Management, 10A,* 127-152.

Scott, B. (1973, March-April) The new industrial estate: Old myths and new realities. *Harvard Business Review, 51*(2), 133-148.

Senn, J. A. (1996, Summer). Capitalizing on electronic commerce: The role of the Internet in electronic markets. *Information Systems Management, 13*(3), 15-24.

Shan, W., Walker, G., & Kogut, B. (1994). Interfirm cooperation and startup innovation in the biotechnology industry. *Strategic Management Journal, 15,* 387-394.

Steer, P., & Cable, J. (1978). Internal organization and profit: An empirical analysis of large UK companies. *Journal of Industrial Economics, 30*(2), 173-211.

Stern, L., & Reve, T. (1980). Distribution channels as political economies: A framework for comparative analysis. *Journal of Marketing, 37,* 39-47.

Stern, R. N. (1979). The development of an interorganizational control network: The case of intercollegiate athletics. *Administrative Science Quarterly, 24,* 242-266.

Stewart, T. A. (1992, May). The search for the organization of tomorrow. *Fortune, 125*(10), 92-98.

Stinchcombe, A. L. (1959). Social structure and organization. In J. G. March (Ed.), *Handbook of organizations* (pp. 142-193). Chicago: Rand McNally.

Teece, D. J. (1981). Internal organization and economic performance: An empirical analysis of the profitability of principal firms. *Journal of Industrial Economics, 30*(2), 173-211.

Teece, D. J. (1987). Profiting from technological innovation: Implications for integration, collaboration, licensing and public policy. In D. J. Teece (Ed.), *The competitive challenge: Strategies for industrial innovation and renewal* (pp. 185-219). Cambridge, MA: Ballinger.

Thanheiser, H. (1973). *The strategy and structure of West German enterprise.* Unpublished doctoral dissertation, Harvard Business School, Boston.

Thompson, J. D. (1967). *Organizations in action: Social science bases of administrative theory.* New York: McGraw-Hill.

Thorelli, H. B. (1986). Networks: Between markets and hierarchies. *Strategic Management Journal, 7,* 37-51.

Tsang, E. W. K. (2000). Transaction cost and resource-based explanations of joint ventures: A comparison and synthesis. *Organization Studies, 21*(1), 215-242.

Tushman, M., & Nadler, G. (1978). Information processing as an integrative concept in organizational design. *Academy of Management Review, 1,* 613-624.

Ulrich, D., & Barney, J. (1984). Perspectives in organizations: Resource dependence, efficiency, and population. *Academy of Management Review, 9,* 471-481.

Uzzi, B. (1996). The sources and consequences of embeddedness for the economic performance of organizations: The network effect. *American Sociological Review, 61,* 674-698.

Uzzi, B. (1997). Social structure and competition in interfirm networks: The paradox of embeddedness. *Administrative Science Quarterly, 42,* 35-67.

Vanhonacker, W. (1997). Entering China: An unconventional approach. *Harvard Business Review, 75*(2), 130-140.

Venkatraman, N., & Bensaou, M. (1996, June 1). Inter-organizational relationships and information techonology: A conceptual synthesis and a research framework. *European Journal of Information Systems, 5*(2), 84.

Volberda, H. W. (1996). Toward the flexible form: How to remain vital in hypercompetitive environments. *Organization Science, 7*(4), 359-374.

Weber, M. (1947). *The theory of social and economic organization.* Glencoe, IL: The Free Press.

Whistler, T. L. (1970). *The impact of computers on organizations.* New York: Praeger Publishers.

Whitt, J. A. (1980). Can capitalists organize themselves? In G.W. Domhoff (Ed.), *Power structure research* (pp. 97-113). Beverly Hills, CA: Sage Publications.

Whittington, R., & Mayer, M. (1997). Beyond or behind the M-form? The structures of European business. In D. O'Neal, H. Thomas, & M. Ghertman (Eds.), *Strategy Structure and Style* (pp. 241-258). Chichester: John Wiley.

Williamson, O. E. (1975). *Markets and hierarchies: Analysis and antitrust implications.* New York: Free Press.

Williamson, O. E. (1985). *The economic institution of capitalism.* New York: Free Press.

Williamson, O. E. (1991). Comparative economic organization: The analysis of discrete structural alternatives. *Administrative Science Quarterly, 36,* 269-296.

Williamson, O. E., & Bhargava, N. (1972). Assessing and classifying the internal structure and control apparatus of the modern corporations. In K. Cowling (Ed.), *Market structure and corporate behavior* (pp. 125-148). London: Macmillan.

Winch, G., Gyllstrom, H., Sauer, F., & Seror-Märklin, S. (1997). The virtual neural business system: A vision for IT support for network form organization. *Management Decision, 35*(1), 40-48.

Woodward, J. (1965). *Industrial organization: Theory and practice.* London: Oxford University Press.

Woodward, J. (Ed.). (1970). *Industrial organization: Behaviour and control.* Oxford: Oxford University Press.

Yates, J. (1989). *Control through communication: The rise of system in American management*. Baltimore, MD: Johns Hopkins University Press.

Yates, J., & Orlikowski, W. (1992). Genres of organizational communication: An approach to studying communication and media. *Academy of Management Review, 17,* 299-326.

Zeffane, R. (1995). The widening scope of inter-organizational networking: Economic sectoral and social dimensions. *Leadership & Organization Development Journal, 16*(4), 26-33.

Zenger, T. R., & Hesterly, W. S. (1997). The disaggregation of corporations: Selective intervention, high-powered incentives, and molecular units. *Organization Science, 8*(3), 209-222.

Zuboff, S. (1988). *In the age of the smart machine: The future of work and power*. New York: Basic Books.

Zucker, L. (1987). Institutional Theories of Organizations. *Annual Review of Sociology, 13,* 443-465.

Endnotes

1 Baum and Dutton (1996) offer an interesting collection of articles in this area.

2 Workflow tools make the task automatically flow to the person or machine to which it corresponds to carry out a particular process or set of interrelated tasks at any moment in time, functioning iteratively until the process is completed. They permit the automation and management of business processes, through the automated coordination, control, and communication of the work to be carried out, by means of the intelligent movement of information.

3 With span of control understood to mean the number of subordinates reporting to a manager.

4 Horizontal specialization refers to the number of tasks undertaken in a particular post. Vertical specialization refers to employees' degree of autonomy and control over their own work.

Chapter II

Intellectual Basis of Research on Alliances: A Bibliometric Study

José Aurelio Medina-Garrido, University of Cádiz, Spain

Antonio Rafael Ramos-Rodriguez, University of Cádiz, Spain

Abstract

The disorder that is apparent in the literature on inter-organizational relationships makes this field ideal for conducting quantitative studies that clarify the conceptual map and complement existing qualitative research. These latter studies only reflect the point of view of a small number of expert researchers, and hence provide a somewhat biased perspective on inter-organizational relationship research. In this context, a bibliometric study should contribute to giving form to the state of the art. In this chapter, a bibliometric study reveals the theoretical fragmentation of the literature on inter-organizational relationships; maps the most relevant studies and theoretical approaches; and highlights the topics that have been addressed most in this field.

Introduction

In chapter one, Medina, Martinez, and Ruiz conducted a review of the research on alliances and other types of inter-organizational relationships. This chapter, similarly to other known works (see García Falcón & Medina Muñoz, 1998; Grandori & Soda, 1995; Gulati, Nohria, & Zaheer, 2000; Oliver & Ebers, 1998; Osborn & Hagedoorn, 1997), attempts to clarify this field of study from perspectives that frequently diverge. Those works are qualitative in nature and reflect the point of view and experience of only a few expert researchers. Thus, they provide a somewhat biased view of the field, although no less valuable for that.

In this context, a bibliometric study could serve to complete the state of the art from a more objective perspective. In this chapter we analyze the most representative citations from a collection of research works on inter-organizational relationships, as well as the frequency with which two of these citations appear in different works together (co-citation), as an indicator of the proximity between them. This may allow us to identify schools of thought in the field. The validity of this technique for exploring the intellectual structure of a scientific discipline has been demonstrated in numerous works (Culnan, O'Reilly, & Chatman, 1990; McCain, 1986; Ramos & Ruiz, 2004; Small, 1973; White & Griffith, 1981; White & McCain, 1998; Ying, Gobinda, & Schubert, 1999).

A graphical representation based on the co-citation of the most influential works in the discipline replaces the criterion of a small group of experts (as occurs in qualitative literature reviews) with that of as many experts as there are works being analyzed. The intention is not to provide a substitute for qualitative analyses of the state of the art, but rather to complement them with a more objective technique. This is considered necessary for relatively unconsolidated fields of research, which, like this one, are highly fragmented (García Falcón & Medina Muñoz, 1998).

Documents with high co-citation frequencies tend to cluster together when they are mapped. These maps can be regarded as a picture of the field in terms of the works of its main authors. They can reveal the evolution of the researchers, show relations between fields, and establish how closely related some disciplines are to others. The maps identify which scientific topics have been investigated most actively, and which individuals, publications, or institutions are currently preeminent in a particular area of research (Garfield, 1986).

This chapter offers two main contributions to the literature. First, it attempts to demonstrate how a bibliometric analysis can contribute to clarifying the state of the art in a little-explored or developing discipline. Second, it aims to shed some light on the research on inter-organizational relationships.

Methodology

Bibliometric techniques have traditionally been classified into two categories, depending on whether they produce activity or relationship indicators. The first provide data about the impact or influence of research contributions, while the second trace the links and interactions between different researchers and research fields, so that the content of the research and its evolution is described (Callon, Courtial, & Penan, 1993). In parallel to this, the current chapter has been carried out in two stages. In the first, we conducted a citation analysis, computing the frequency of citation of the bibliographic references used in all the articles of a representative sample of the research on inter-organizational relationships that we shall henceforth call citing sample. In the second stage, we used co-citation analysis to trace the relations between the most influential works and identify schools of thought and research topics active during the period analyzed.

The citing sample we used consisted of research articles on inter-organizational relationships published in a collection of journals that were selected by a panel of experts from among the most influential research journals in the management area[1] (Tahai & Meyer, 1999).

Choosing the article, rather than other academic sources such as books, doctoral theses, or scientific congress records, is justified since articles can be regarded as certified knowledge, in other words knowledge that has been submitted to the critical review of fellow researchers and that has overcome their objections (Callon et al., 1993).

Given the nature of the citation analysis of this chapter, the primary source of information was the Social Science Citation Index (SSCI). This is an index of citations run by the U.S. Institute for Scientific Information (ISI), which has compiled the content of around 4,100 journals of worldwide distribution since 1972. Despite the long time period of this database, the period studied in the current chapter runs from January 1992 to February 2001, because of our limited access via the Centro Informático Científico de Andalucía (CICA).[2] But given the novelty of the topic dealt with here and the size of the citing sample, the period chosen can be considered to be representative of the current research on inter-organizational relationships.

After first collecting all the articles published during the period of analysis and filtering them on the basis of key terms from the discipline, we reviewed each article by hand to reject those bearing no relation with the field of study. For example, using the keyword *networks* in the search resulted in articles on computing networks as well as those dealing with strategic networks of firms. The first set of articles had to be excluded. In this way, the final citing sample consisted of 319 articles directly related with the field of interest.

Subsequently, we compiled a database with all the bibliographic references used in the articles of the citing sample. In order to ensure the reliability of the subsequent citation counts, this database was painstakingly checked by hand to correct spelling mistakes or errors in the coding of the citations, as well as to ensure that only the first editions of the books were included.

Co-citation analysis initially used the work cited as the unit of analysis, but difficulties in handling this data led the methodology to evolve towards the citation analysis of any works of a particular author, that is, the author's complete works. These two methods have similar approaches, but the second suffers from a number of important limitations, and its results are difficult to interpret, above all, in areas as multi-disciplinary as management, where an author can make important contributions to the discipline from different perspectives. Hence, in this chapter we have used the work or document cited as the unit of analysis.

On the basis of the stages that normally make up an author co-citation analysis, we now describe in more detail the stages followed in our document co-citation analysis. This technique is based on counting, within a set of articles representative of the research in a particular scientific discipline, the number of articles that cite a particular pair of documents. White and Griffith (1981); McCain (1990); and White and McCain (1998) describe this procedure applied to author co-citation analysis in detail. Its validity for exploring the intellectual structure of a scientific discipline has been amply demonstrated in numerous studies (Culnan et al., 1990; McCain, 1986; White & Griffith, 1981; White & McCain, 1998; Small, 1973; Ying et al., 1999).

In order to build the map of the intellectual structure of the research on inter-organizational relationships we selected the 26 most influential works identified in the previous citation frequency count, which corresponds to a citation threshold of at least 10% of the works in the citing sample.

The maps are built by calculating Pearson's correlation coefficient between the co-citation profiles of each pair from the 26 most cited works. To build this measure, we calculate the distribution of the citation frequencies of all the works included in the bibliographic references in the articles of the citing sample. From there, all the possible pairs among the 26 most cited documents are taken and a score is recorded of how many articles cite each pair.

The counts obtained are arranged on a square, symmetric matrix, whose main diagonal is left undefined. Subsequently, this matrix of raw co-citation frequencies is converted into a correlation matrix whose coefficients reflect the measure of proximity required.

For this, in practically all works involving author co-citation analysis Pearson's correlation coefficient is used as the similarity measure (McCain, 1990; White & Griffith, 1981; White & McCain, 1998). In this study we carry out a co-citation analysis of documents, not authors, we have also opted for this measure, for at least two reasons. First, for a given pair of documents Pearson's correlation coefficient is a measure not of the frequency that this pair of documents is cited (the raw co-citation frequency), but rather the similarity between their co-citation profiles and those of the rest of the works considered. Thus, two works that are always co-cited with a third work, but rarely with others, will have strong positive correlation and we can say that they are perceived as being related or similar in some sense by the citing sample. Second, the correlation coefficient also gets round differences of scale between those works that are cited very often and those with similar citation profiles that are less frequently cited (Kerlinger, 1973; White & McCain, 1998).

There are two ways of treating the main diagonal when calculating the correlation coefficients. The first (White & Griffith, 1981) calculates the result of summing the three highest scores and dividing by two, which can indicate in a general way the relative importance of a given work in the field. The other option (McCain, 1990) is to treat it as missing data and apply the criterion of eliminating these two cases from the calculation of the correlation coefficients (pairwise delete).

In the current work we adopted the second technique, since no significant differences were found in the resulting configurations after carrying out both analyses. For this reason, and because it is simpler, we opted to ignore the scores on the main diagonal in the calculation of the correlation coefficients between the pairs of documents.

The next step in the process had the aim of computing the coordinates of the items analyzed in a sufficiently reduced space for their subsequent graphical representation and interpretation. The statistical technique chosen for this was multi-dimensional scaling, (MDS). MDS analysis requires the use of a similarity matrix between objects as input. This procedure results in a map produced from the similarity matrix between the elements analyzed, so that the structure underlying the set of items can be studied. In a co-citation analysis there are two main uses of the MDS procedure: on the one hand, it provides information about the co-citation relationships, and on the other, it helps to identify the underlying relationships observed between the positions (McCain, 1990).

Documents with similar co-citation profiles tend to cluster together in the new configuration. Thus, those works that are closely related to others tend to occupy a central position in the intellectual space, while those with only loose relations tend to be located on the periphery. In this way, it is easy to determine whether schools of thought or other intellectual groups are central or peripheral.

The result of the MDS is a table with the coordinates for each document plotted on a two-dimensional plane. Its main aim is to collate the maximum amount of information from the original data in only two or three dimensions, that is, to reduce the spatial dimension. This simplification inevitably distorts the original distances and cannot fully explain all the variance that appears in the similarity matrix.

The statistical package used measures the distortion with a goodness-of-fit index labeled stress,[3] which provides an approximate measure of the difference between the distances between points in the initial configuration and those obtained in the final configuration. The evaluation of the goodness-of-fit can be complemented with a graphical representation (Sheppard's diagram), where the original distances are plotted on the X-axis and the distances in the new configuration on the Y-axis, so that with the appropriate scale a cloud of points clustering around the diagonal would be indicative of goodness-of-fit. On the other hand, the goodness-of-fit directly depends on the number of units analyzed, so that the more units that are mapped, the poorer the goodness-of-fit. This was an additional reason for the decision to limit the analysis to the 26 most cited articles, thereby ensuring an adequate stress.

With the coordinates calculated for each document, the new configurations obtained were plotted on a map. Briefly, the distances observed in the map give some idea of the similarity between any two works.

Results of Bibliometric Study

Co-Citation Map

Figure 1 shows the map of documents resulting from the technique described in the previous sections. In this map, the numbers equate to the order of citation, with number one being the work most cited by all the articles in the citing sample. The position of the number in question (not that of the label containing author and year) represents the location of the corresponding work in this two-dimensional map. Alongside each number, the first author and the year of publication are mentioned. Where there is more than one author only the first of these is given.[4]

As has already been mentioned, in a consolidated field of research such a map can be expected to reveal groups of works characterizing the different schools or disciplines, addressing the theme in question, and also the most important topics in the area. However, and in line with the arguments defended by García Falcón and Medina Muñoz (1998), the inter-organizational relationship's field is examined by a fertile but fragmented set of literature. In this respect, it is difficult for groups to take shape, and rather than schools, we find a mixture of theoretical approaches with topics from the field of study.

Figure 1. Intellectual basis of inter-organizational relationship research

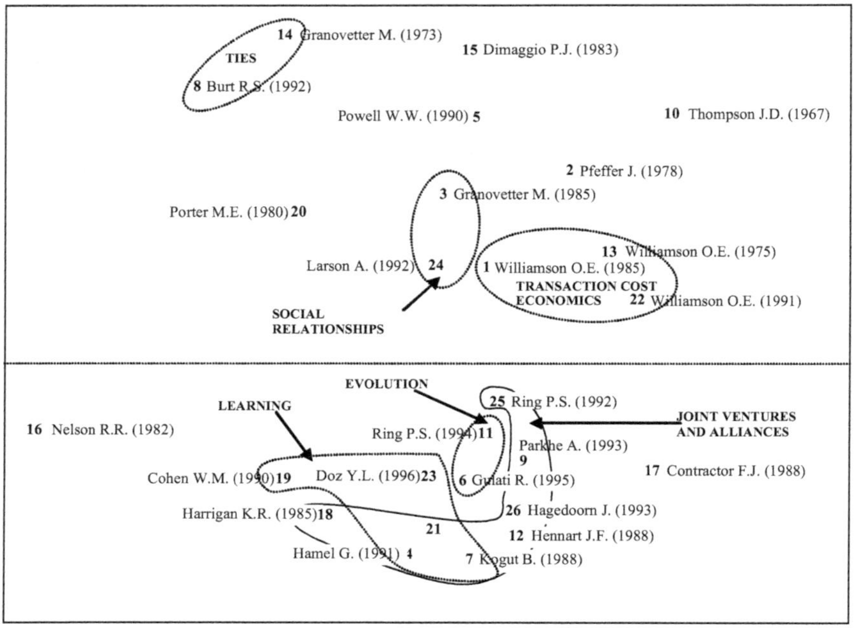

We now conduct a descriptive analysis of the information represented in the map. We analyze the groups of works detected, the proximity or remoteness between groups, and the correlation between works.

Groups of Works

A first approximation allows us to see two broad groups within the already mentioned fragmentation, separated by a discontinuous line in Figure 1. In general terms, the group in the top half of the figure consists of works dedicated to theoretical developments, while the bottom group focuses on topics from the field of inter-organizational relationships.

More specifically, the top group contains those works that constitute the conceptual frameworks used in research on inter-organizational relationships. The distances between the different works within this group allow us to deduce that although there is indeed a strong multi-disciplinarity (García Falcón & Medina Muñoz, 1998), the eclectic approach that is required has not been commonly adopted (Tsang, 2000).

We can distinguish three subgroups within this top group, in terms of both the content of the works and the distances represented in the map, with the remaining works lacking any viable grouping. The first subgroup consists of Burt (1992) and Granovetter (1973), and analyzes the nature of the ties between organizations. The importance of the relationships and the social dimension allows us to group Larson (1992) and Granovetter (1985) into a second subgroup. The third subgroup represents the contribution of transaction cost economics (TCE) to the field of inter-organizational relationships; it is based almost exclusively on the work of Williamson (1975, 1985, 1991).

The remaining authors, despite having been cited very frequently, do not share similar co-citation profiles, so they fail to consolidate conceptual schools in the field. Thus, the following work has been widely cited: Porter (1980), with his contribution of the structure-conduct-performance paradigm to industrial organization; Powell (1990), who regards inter-organizational relationships as a different organizational form from that of the market or hierarchy; the institutional approach of the work of Dimaggio and Powell (1983); Pfeffer and Salancik's (1978) resource dependence theory, which has been used to analyze power struggles within inter-organizational relationships; and Thompson's (1967) work, which is a pioneering study on corporate strategies and the formation of coalitions.

We should note that Porter's work does not directly relate to the work on inter-organizational relationships. Its high citation count in the sample analyzed can be explained by the number of times it is cited when all the works in, for example, the *Strategic Management Journal*, are analyzed (see Ramos & Ruiz, 2004). This effect could be called citation by inertia, and it is due to the tendency of authors to inevitably cite, in their works, certain articles that have had a strong impact and that are highly prestigious in the area of knowledge, regardless of their direct relation with the topic under analysis.

The second group, below the discontinuous line in Figure 1, presents a mix of works that could be labeled, depending on the case, as work addressing specific topics or alternatively as work of a more empirical nature. The first thing we note from a simple glance at this part of the map is that the different documents are closer together than in the first group. This closeness demonstrates a greater co-citation index, except in two cases: Nelson and Winter's (1982) work on evolutionary theory,[5] and Contractor and Lorange's (1988) work on cooperative strategies in international business.

Like in the previous case, different subgroups can be distinguished here too. A first subgroup seems to contain those articles that have contributed important bases for the discipline in two of the most frequently cited inter-organizational relationship types: joint ventures and alliances. Thus, other forms of relationships such as franchises, licensing agreements, subcontracting, and so forth, receive less attention from the most influential articles in the field.

With regards to contributions dealing with joint ventures, in this subgroup we find the work of Kogut (1988), which considers the learning capacity of joint ventures, as well as their capacity to raise market power; Hennart (1988) applies transaction cost economics to equity joint ventures, and considers that this type of collaboration can be explained as a way of getting round inefficient markets for intermediate inputs; and Harrigan (1985) examines when firms will prefer to cooperate through joint ventures rather than through other forms of cooperation.

As far as alliances are concerned, the following works are also located in that subgroup: Hamel (1991), which deals with the determinants and asymmetries of learning between alliance members, arguing that the value of an alliance can come from the benefits to be gained from the successful execution of joint tasks or from internalizing other members' capabilities; Hamel, Doz, and Prahalad (1989) analyze the factors on which the success of alliances depend, such as the convergence between strategic objectives, divergence in competitive objectives, size, market power, or inter-organizational learning; Parkhe (1993) studies alliance structure and its connection with performance, making use of game theory and transaction cost economics; Ring and Van de Ven (1992) study four forms of alliance governance in function of the risk and the need for trust; and finally, Hagedoorn (1993) attempts to clarify the motives driving firms to cooperate in their efforts to innovate.

A second subgroup contains those works dealing with one of the topics that have most interested scholars in this literature: learning. Some of the works overlap with the previous subgroup, namely: Hamel (1991); Kogut (1988); and Hamel et al. (1989). Apart from these, in this group we find Cohen and Levinthal's (1990) work, which deals with the concept of absorptive capacity, referring to a company's ability to recognize the value of external information, assimilate it, and apply it for commercial purposes; and Doz (1996), who analyzes learning in strategic alliances along dimensions such as the environment, task, processes, abilities, and objectives, and predicting how different learning processes can determine the success or failure of an alliance.

A final subgroup consisting of only two works can be identified. In this subgroup the focus is on the evolutionary process of the alliance. It contains the work of Ring and Van de Ven (1994), who present a processual framework explaining how and why inter-organizational relationships emerge, evolve, and dissolve; and that of Gulati (1995), which attempts to improve on the transaction cost approach, which, according to this author, incorrectly treats each transaction as independent and ignores the role of the trust that emerges when firms engage in repeated alliances. We should mention the strong impact of this latter work, since despite being one of the most recent—1995—it occupies sixth place among the most cited works.

Location of Groups

As we mentioned in the previous section, two large areas can be distinguished in Figure 1, divided by a horizontal discontinuous line. The works in the top half of the figure represent some of the conceptual frameworks that have been most used, from, at times, diametrically opposed perspectives, for example, social networks (Granovetter, 1973, 1985; Larson, 1992), the governance mechanisms employed (Williamson, 1975, 1985, 1991), industrial organization (Porter, 1980), institutional theory (Dimaggio & Powell, 1983), resource dependence theory (Pfeffer & Salancik, 1978), or from a strategic management approach, the pioneering concepts of Thompson (1967) on cooperative strategies. The works in the bottom half of the figure, as we have also mentioned, focus on the study of specific topics relating to learning and knowledge management, the formation and evolution of alliances and joint ventures, and so forth.

The previous theoretical frameworks and topics treated have led to the formation of different groups of works. However, it is also interesting to analyze the position of some groups compared to others. Just as the proximity between two works indicates a similar co-citation profile, and hence the possible formation of schools, the proximity between groups indicates a close relation between the topics dealt with or theoretical approaches used.

In this respect, the theoretical works in the top part of the figure show, judging by their distances, that they tend to be cited together rarely, except for the case of social networks (Granovetter, 1985; Larson, 1992) and transaction cost economics (Williamson, 1975, 1985, 1991). There are two reasons for the high co-citation of these works. The first lies in the importance of some social topics, for example, trust, or the reputation of individuals (which is extendable to organizations) in decisions regarding the most appropriate form of governance, and in the emergence of otherwise opportunistic behaviors among the members of a relationship. The second reason, rather than complementarity shows the antagonism of works based on the concept of social networks that cite works on transaction cost economics in a critical way, because the latter consider the transactions between firms to be discrete, rather than continuous and founded on enduring relationships and the trust between individuals.

With regards to these two groups—that of the transaction costs and that of the social networks—we should stress their centrality, a clear sign of their important role as an intellectual basis for the whole discipline. These two groups support diverse and very different topics in the field of inter-organizational relationships. They are frequently referred to for making comparisons with other alternative theoretical approaches, which occurs very often due to the discipline's need for an eclectic approach (Tsang, 2000).

Within the first area, the group dedicated to ties in inter-organizational relationships (Burt, 1992; Granovetter, 1973) is located in a peripheral position diametrically opposed to the contributions on transaction cost economics, and, moreover, very distant from all the articles in the second area. This contrasts with its high citation frequency, which could be due to the fact that it is a group without a fixed pattern of citation, which is cited indistinctively with other works with which it fails to form a school.

With regards to the second area, dedicated to more specific topics in the area of inter-organizational relationships, we observe how the works dealing with the topic of joint ventures (Harrigan, 1985; Hennart, 1988; Kogut, 1988) and alliances (Hagedoorn, 1993; Hamel, 1991; Hamel et al., 1989; Parkhe, 1993; Ring and Van de Ven, 1992) have fused into a single subgroup, when they could have been represented separately. However, the close relation between these works has led to them being considered jointly.

This group, on the other hand, is cited very frequently in the works focusing on learning and knowledge management (Cohen & Levinthal, 1990; Doz, 1996; Hamel, 1991; Hamel et al., 1989; Kogut, 1988). This is demonstrated by the proximity between these two subgroups, as well as their overlap in three articles (Hamel, 1991; Hamel et al., 1989; Kogut, 1988).

The group dedicated to the management of relationships (Gulati, 1995; Ring & Van de Ven, 1994) also demonstrates, by its proximity, a direct relation with the two previous groups. In this case it is again clear that they form part of the same intellectual structure, generating a similar literature in terms of the citations used.

Correlation Matrix

Analyzing the correlation matrix allows us to see which works are co-cited most and least often. It is particularly interesting to note cases on the frontier of groups, "borderline" cases in White and McCain's (1998) terminology. Two close-lying borderline works are highly correlated but belong to distinct groups because of the different topics they deal with. These borderline works act as bridges between the different themes addressed in the discipline.

Coinciding with the conclusions of previous sections, when we select pairs of works with a correlation exceeding 0.9, we can observe the high interrelation between the groups of works dealing with joint ventures and alliances, evolution of the inter-organizational relationship, and learning. In order of correlation, the most correlated works are: Kogut (1988) and Hamel et al. (1989); Hamel (1991) and Hamel et al. (1989); Hennart (1988) and Hagedoorn (1993); Ring and Van de Ven (1992) and Ring and Van de Ven (1994); Kogut (1988) and Hennart (1988); and Gulati (1995) and Doz (1996). The works of Kogut (1988) and Hamel et al. (1989), and Hamel (1991) and Hamel et al. (1989) represent the intersection between the group focusing on learning in collaborations and that of joint ventures and alliances. Kogut (1988) and Hennart (1988) are also highly correlated, although perhaps these works should not be regarded as borderline, since Kogut (1988) forms part of both groups. The works of Ring and Van de Ven (1992) and Ring and Van de Ven (1994) constitute a bridge between the group interested in evolution and the one examining joint ventures and alliances, similarly to what occurs with the work of Gulati (1995) and Doz (1996), which link works on learning with works on the evolution of inter-organizational relationships.

In short, despite the fact that some of the previous works are contained within different groups, the existence of high correlations between them demonstrates that these separations are somewhat artificial, and that these works could in fact belong to a single large group labeled characteristics, evolution, and learning in inter-organizational relationships. Once again we are struck by the difficulty of detecting schools in this field; other bibliometric research has detected schools more easily in more consolidated and defined research disciplines (e.g., White & McCain, 1998, for information sciences).

With regards to less correlated works, the following stand out, in this order: Hamel (1991) and DiMaggio and Powell (1983); Hamel et al. (1989) and DiMaggio and Powell (1983); Hamel (1991) and Granovetter (1973); Kogut (1988) and DiMaggio and Powell (1983); and Hagedoorn (1993) and Granovetter (1973).

The extreme and peripheral location in the map of the work of DiMaggio and Powell (1983) and Granovetter (1973) already suggested that there would be no clear pattern of co-citation with the majority of works in the study, despite the frequency with which these works are cited, and analyzing the correlation matrix confirms this fact, particularly with two groups of works. On the one hand, the works on learning in joint ventures or in alliances (Hamel, 1991; Hamel et al., 1989; Kogut, 1988) seem to have little to do with a possible institutional explanation of the phenomenon (the framework established by DiMaggio & Powell, 1983). On the other hand, the strength of weak ties as an important element in the structuring of social networks (Granovetter, 1973) appears to bear little relation, at least according to our citing sample, with the learning processes taking place in alliances (Hamel, 1991) or with the motives driving firms to cooperate in their efforts to innovate (Hagedoorn, 1993).

Limitations

Conducting a bibliometric study to determine the state of the art of a particular area—specifically the literature on inter-organizational relationships—has its limitations. First, the documents analyzed as the source of the bibliometric study are just a small part of the research activity on inter-organizational relationships, so that the conclusions drawn have limited application. But in view of the scientific journals analyzed, the citing sample selected here is sufficiently representative of the literature in the research area of management. Thus, the limitations of this sampling are similar to those of any other statistical sampling.

Second, it is impossible to distinguish whether articles are citing works to criticize them or rather to employ them as part of their reasoning. It is true, though, as López (1996) points out, when an author's work lacks quality it is simply not cited.

In any case, the possible applications of the tool presented here have great potential for research in management. Bibliometrics can be applied to a wide variety of topics, above all the most recent ones, such as inter-organizational relationships, for which we still lack extensive literature reviews. Thus, this type of study may be useful for identifying the state of the art and detecting the most influential authors, journals, and works.

Finally, this type of study is useful for assessing the inter-organizational relationship research in an objective way, complementing—but not substituting for—the traditional qualitative methods of the literature review. It can be used as a tool for identifying the authors, works, and journals most consumed by researchers in the discipline, as well as the relational links between them. Hence, it may provide researchers with guidelines for identifying the relevant literature in a particular research area, building an intellectual map of the topic, and allowing them to gain an objective view of the field based on the behavior of the leading authors themselves.

Conclusions

This chapter has aimed to shed some light on the research in the literature on inter-organizational relationships. This discipline is highly fragmented (García Falcón & Medina Muñoz, 1998), and there are few studies of the state of the art (García Falcón & Medina Muñoz, 1998; Grandori & Soda, 1995; Gulati et al., 2000; Oliver & Ebers, 1998; Osborn & Hagedoorn, 1997). The ones that do exist are incoherent in part of their content and offer a biased and partial view, influenced by the profile of their authors.

In this respect, bibliometrics offers an objective analysis of the opinions of a large number of authors with diverging tendencies and interests in different topics in the area. A significant number of works have confirmed the validity of this technique for exploring the intellectual structure of a scientific discipline (Culnan et al., 1990; McCain, 1986; Ramos & Ruiz, 2004; Small, 1973; White & Griffith, 1981; White & McCain, 1998; Ying et al., 1999).

The graphical representation of the co-citation profiles of the different works reveals a scarcity of schools in the field of inter-organizational relationships, as well as their fragmentation. This contradicts the findings of some studies on the state of the art based on a qualitative perception of the discipline.

Among the schools that can be detected, we should mention the ones based on transaction cost economics (Williamson, 1975, 1985, 1991), or social networks (Granovetter, 1985; Larson, 1992), and the group of works analyzing the nature of the ties in both social and business networks (Burt, 1992; Granovetter, 1973). Also significant are other works dealing with theoretical frameworks that do not manage to consolidate schools in the discipline, but are cited and used in numerous works, namely: Porter's (1980) work, with its contribution of the structure-conduct-performance paradigm, which is the object of strong citation inertia throughout the area of management; Powell (1990) who defines inter-organizational relationships as a form of organization distinct from that of the market or the hierarchy; Dimaggio and Powell's (1983) institutional approach; Pfeffer and Salancik's (1978) resource dependence theory; and Thompson's (1967) pioneering work on cooperative strategies and coalition formation.

With regards to the groups of works focusing on specific topics in the discipline, three groups, which are closely related—and in fact overlap—stand out. This overlap demonstrates the absence of polarization in the intellectual basis underlying the treatment of the key topics in the discipline. Thus, a large part of the articles analyzed in the citing sample come from a common stream made up of works on the topic of learning and knowledge management in inter-organizational relationships (Cohen & Levinthal, 1990; Doz, 1996; Hamel, 1991; Hamel et al., 1989; Kogut, 1988), the evolution of these relationships (Gulati, 1995; Ring & Van de Ven, 1994), and the nature and characteristics of joint ventures and alliances (Hagedoorn, 1993; Hamel, 1991; Hamel et al., 1989; Harrigan, 1985; Hennart, 1988; Kogut, 1988; Parkhe, 1993; Ring & Van de Ven, 1992).

Although the citing sample can be regarded as sufficiently representative, in order to obtain a more precise idea of the structure of the discipline the threshold of citation could be reduced to increase the number of cited works subject to analysis. In this respect, we carried out preliminary tests with the 85 most cited authors. The resulting map demonstrated the growing importance of new schools, for example, that of the resource-based view of the firm. However, with this expanded map the strong heterogeneity of the articles, the numerous topics analyzed, and the consequent difficulty in grouping the documents, on the one hand, and the poorer goodness-of-fit, on the other, make it doubtful about the wisdom of extending the current analysis from 26 to 85 articles.

In this respect, a future line of research that may be viable would be to represent the main authors rather than specific works, as has been mentioned in the methodology. This could reveal the formation of schools more clearly, but it has the drawback that these authors could possess a multi-disciplinary work that would distort the study to some extent.

Finally, we should reiterate that this quantitative methodology is not meant to substitute for qualitative studies on the state of the art. Rather, the methodology presented here can complement these studies in the case of a fragmented discipline that has yet to be fully defined.

References

Barney, J. B. (1991). Firm resources and sustained competitive advantage. *Journal of Management, 17*(1), 99-120.

Burt, R. S. (1992). *Structural holes: The social structure of competition.* Cambridge, MA: Harvard University Press.

Callon, M., Courtial, J. P., & Penan, H. (1993). *Cienciometría. La medición de la actividad científica: De la bibliometría a la vigilancia tecnológica.* Ediciones Trea, S.L. (*Sciencemetric. The measurement of the scientific activity: from bibliometric to technological monitoring.* Gijon, Spain: Trea Publishing.)

Cohen, W., & Levinthal, D. A. (1990). Absortive capacity: A new prespective on learning and innovation. *Administrative Science Quarterly, 35,* 128-152.

Contractor, F. J., & Lorange, P. (1988). *Cooperative strategies in international business.* Lexington: Lexington Books.

Culnan, M. J., O'Reilly, C. A., & Chatman, J. A. (1990). Intellectual structure of research in organizational behavior, 1972-1984: A cocitation analysis. *Journal of the American Society for Information Science, 41*(6), 453-458.

Dierickx, I., & Cool, K. (1989). Asset stock accumulation and sustainability of competitive advantage. *Management Science, 35*(12), 1504-1511.

Dimaggio, P. J., & Powell, W. W. (1983). The iron cage revisited: Institutional isomorphism and collective rationality in organizational fields. *American Sociological Review, 48,* 147-160.

Doz, Y. (1996). The evolution of cooperation in strategic alliances: Initial conditions or learning processes? *Strategic Management Journal, 17,* 55-83.

Eisenhardt, K. M. (1996). Resource-based view of strategic alliance formation: Strategic and social effects in entrepreneurial firms. *Organization Science, 7*(2), 136-150.

García Falcón, J. M., & Medina Muñoz, D. R. (1998). Enfoques teóricos sobre las relaciones interorganizativas: Una revisión comparativa. (Approaches to interorganizational relationship: A comparative revision.) *Revista Europea de Dirección y Economía de la Empresa, 7*(3), 9-30.

Garfield, E. (1986). *Essay of an information scientist: Towards scientography* (Vol. 9). Philadelphia: ISI Press.

Grandori, A., & Soda, G. (1995). Inter-firm network: Antecedents, mechanisms and forms. *Organization Studies, 16*(2), 183-214.

Granovetter, M. (1973). The strength of weak ties. *American Journal of Sociology, 78,* 1360-1380.

Granovetter, M. (1985). Economic action and social structure: The problem of embeddedness. *American Journal of Sociology, 91,* 481-510.

Gulati, R. (1995). Does familiarity breed trust? The implications of repeated ties for contractual choice in alliances. *Academy of Management Journal, 38,* 85-112.

Gulati, R., Nohria, N., & Zaheer, A. (2000). Strategic networks. *Strategic Management Journal, 21,* 203-215.

Hagedoorn, J. (1993). Understanding the rationale of strategic technology partnering: Inter-organizational modes of cooperation and sectoral differences. *Strategic Management Journal, 14,* 371-385.

Hamel, G. (1991). Competition for competence and inter-partner learning within international strategic alliances. *Strategic Management Journal, 12,* 83-104.

Hamel, G., Doz, Y., & Prahalad, C. (1989). Collaborate with your competitors and win. *Harvard Business Review, 67,* 133-139.

Harrigan, K. R. (1985). *Strategies for joint ventures.* London: Lexington.

Hennart, J. F. (1988). A transaction costs theory of equity joint ventures. *Strategic Management Journal, 9*(4), 361-374.

Kerlinger, F. N. (1973). *Foundations of Behavioral research.* New York: Holt, Rinehart & Winston.

Kogut, B. (1988). Joint ventures: Theoretical and empirical perspectives. *Strategic Management Journal, 9*(4), 319-332.

Larson, A. (1992). Network dyads in entrepreneurial settings: A study of the governance of exchange processes. *Administrative Science Quarterly, 37,* 76-104.

López, P. (1996). *Introducción a la bibliometría.* (*Introduction to bibliometric.*) Madrid, Spain: Promolibro.

McCain, K. W. (1986). Cocited author mapping as a valid representation of intellectual structure. *Journal of the American Society for Information Science, 37*(3), 111-122.

McCain, K. W. (1990). Mapping authors in intellectual space: A technical overview. *Journal of the American Society for Information Science, 41*(6), 433-443.

Nelson, R. R., & Winter, S. (1982). *An evolutionary theory of economic change.* Cambridge, MA: Harvard University Press.

Oliver A. L., & Ebers, M. (1998). Networking network studies: An analysis of conceptual configurations in the study of inter-organizational relationships. *Organization Studies, 19*(4), 549-583.

Osborn, R. N., & Hagedoorn, J. (1997). The institutionalization and evolutionary dynamics of interorganizational alliances and networks. *Academy of Management Journal, 40*(2), 261-278.

Parkhe, A. (1993). Strategic alliance structuring: A game theoretic and transaction cost examination of interfirm cooperation. *Academy of Management Journal, 36,* 794-829.

Penrose, E. T. (1959). *The theory of the grow of the firm.* New York: Wiley.

Pfeffer, J., & Salancik, G. R. (1978). *The external control of organizations: A resource dependence perspective.* New York: Harper & Row.

Porter, M. E. (1980). *Competitive strategy.* New York: Free Press.

Powell, W. W. (1990). Neither market nor hierarchy: Network forms of organization. In B. M. Staw & L. L. Cummings (Eds.), *Research in organizational behavior* (Vol. 12, 295-336). Greenwich, CT: JAI Press.

Ramos, A. R., & Ruiz, J. (2004). Changes in the intellectual structure of strategic management research: A bibliometric study of the strategic management journal, 1980-2000. *Strategic Management Journal, 25,* 981-1004.

Ring, P. S., & Van de Ven, A. H. (1992). Structuring cooperative relationships between organizations. *Strategic Management Journal, 13,* 483-498.

Ring, P. S., & Van de Ven, A. H. (1994). Developmental processes of cooperative interorganizational relationships. *Academy of Management Review, 19*(1), 90-118.

Small, H. G. (1973). Co-citation in the scientific literature: A measure of the relationship between two documents. *Journal of the American Society for Information Science, 24,* 265-269.

Tahai, A., & Meyer, M. J. (1999). A revealed preference study of management journals' direct influences. *Strategic Management Journal, 20,* 279-292.

Thompson, J. D. (1967). *Organizations in action: Social science bases of administrative theory.* New York: McGraw-Hill.

Tsang, E. W. K. (2000). Transaction cost and resource-based explanations of joint ventures: A comparison and synthesis. *Organization Studies, 21*(1), 215-242.

Wernerfelt, B. (1984). A resource-based view of the firm. *Strategic Management Journal, 5,* 171-180.

White, D. H., & McCain, K. W. (1998). Visualizing a discipline: An author cocitation analysis of information science, 1972-1995. *Journal of the American Society for Information Science, 49*(4), 327-355.

White, H. D., & Griffith, B. C. (1981). Author cocitation: A literature measure of intellectual structure. *Journal of the American Society for Information Science, 32,* 163-171.

Williamson, O. E. (1975). *Markets and hierarchies.* New York: Free Press.

Williamson, O. E. (1985). *The economic institution of capitalism.* New York: Free Press.

Williamson, O. E. (1991). Comparative economic organization: The analysis of discrete structural alternatives. *Administrative Science Quarterly, 36,* 269-296.

Ying, D., Gobinda, C., & Shubert, F. (1999). Mapping the intellectual structure of information retrieval studies: An author co-citation analysis, 1987-1997. *Journal of Information Science, 25*(1), 67-78.

Endnotes

1 *Academy of Management Journal, Academy of Management Review, Administrative Science Quarterly, California Management Review, Harvard Business Review, Journal of Management, Journal of Management Studies, Management Science, Organization Science, Organization Studies, Sloan Management Review, and Strategic Management Journal.*

2 Centro Informático Científico de Andalucía (CICA) can be translated as "Scientific and Computing Center of Analusia."

3 The raw stress value of a configuration is defined by: $stress = \sum \left[d_{ij} - f(\delta_{ij}) \right]^2$ Where δ_{ij} denotes the reproduced distances, given the respective number of dimensions, and äij denotes the input data (i.e., observed distances). The expression $f(\delta_{ij})$ indicates a non-metric, monotone transformation of the observed input data (distances). Thus, the smaller the stress value, the better the fit of the reproduced distance matrix to the observed distance matrix.

4 Mentioning only the first author is a limitation that is a consequence of using the SSCI database. However, the references at the end of this work contain the names of all the authors of each work.

5 Although Nelson and Winter's work is isolated on this map, in the map built to represent the 85 most cited works (on which we decided not to work in view of the dispersion and fragmentation of the documents, not to mention the worse fit, measured by the stress), it appears perfectly grouped together with the work of Barney (1991); Wernerfelt (1984); Dierickx and Cool (1989); Penrose (1959); and Eisenhardt (1996). Thus, a resource-based view of the firm is taking shape in the treatment of alliances. This stream consolidated much later than the rest of the theoretical work cited, so that it has not had time to exert a greater impact on the current map. Nelson and Winter (1982) contribute, among other things, with their concept of organizational routines to this line of research.

Chapter III

Exploring Trust Building Mechanisms in Global B2B Electronic Markets

Ana Rosa del Águila-Obra, University of Málaga, Spain

Antonio Padilla-Meléndez, University of Málaga, Spain

Abstract

There have been numerous studies about business-to-business (B2B) electronic commerce and market structure, most of them analyzing the relationships between buyers and sellers and the role of intermediaries. This research is based mainly on earlier papers about the role of information technology (IT) and electronic communications networks in the companies' relationships. The use of these computerized inter-organizational networks leads to lower transaction costs, which encourages the development of electronic markets, where there is a severe price competition and greater buyer choice. In this chapter, we analyze the development of global B2B electronic markets and if these markets are becoming a way of improving trust between organizations at an international level, increasing, therefore, the inter-organizational cooperation among them. Based on the literature review and on the analysis of some global electronic markets, we define and describe these platforms, stressing the strategic role of each of the principal participating actors. Furthermore, we propose a model to explain the trust-related sources of competitive advantage for the new intermediaries in electronic markets and compare those with the off-line market's characteristics. In addition, some propositions related with the trust-building mechanisms are defined.

Introduction

On the Internet, businesses can collaborate with other companies through third-party structures or institution-based trust mechanisms, known as B2B Internet-based electronic marketplaces or Global b2b Electronic Marketplaces (Gb2bEM). As global electronic markets are spreading, trust is becoming an important issue in these global online environments (Ratnasingham, 1998). Furthermore, it is becoming difficult for the involved partners to build trust in global markets without face-to-face interaction (Walther, 1995).

In these markets, buyers interact, sometimes, with unknown sellers and vice versa (Pavlou & Gefen, 2004). Normally, in these transactions, buyers and sellers use an intermediary, Gb2bEM, that ensures the transactions with trust-building mechanisms, such as online feedback, monitoring, legal bonds, accreditation, and cooperative norms (Ba & Pavlou, 2002; Dellarocas, 2003; Pavlou, 2002). In addition, Pavlou and Gefen (2004) argue that an electronic market will have success only if it considers the design of adequate online trust-building mechanisms. In other words, in online interchanges, organizational buyers and sellers do the transaction by using Gb2bEM as intermediaries, because of its mediation function. They collect information about the agents involved on the market, inform of experiences on the market, and punish the agents with a non-desirable behaviour, even expelling them from the market. In this sense, our research questions for this chapter are twofold: Firstly, how do the Gb2bEM generate trust between buyers and sellers on online global environments? Secondly, does the presence of online trust building mechanisms influence the Gb2bEM success?

The rest of the chapter is organized as follows. Firstly, the role of intermediaries and the Internet will be analyzed. Secondly, Gb2bEM, as Internet-based intermediaries will be described, both from the Internet business models (describing some examples of Gb2bEM) and from the inter-organizational trust perspective. Finally, the main elements and relations related to trust building on electronic markets are considered in a model to explain the Gb2bEM success.

Literature Review

Intermediaries and the Internet

The impact of the Internet on businesses can be studied from both the industry and the firm's perspective. From an industry perspective, the Internet has positive and negative effects on sellers and buyers (Benjamin & Wigand, 1995, 1997). The positive effects for sellers are mainly a reduction of the distance between buyer and supplier and an increase in the sellers' change costs. On the other hand, a negative effect is that the Internet normally reduces the sellers' profit margin. For buyers, the Internet enhances their access to catalogues of products from multiple sellers, online order processing, and in many cases, discounted prices, such as in the airline industry (Buhalis, 2004).

The Internet creates new industries, reconfigures others, and has a direct impact on companies, buyers, sellers, distributors, and potential new entrants (Porter, 2001). Furthermore, it has been argued that with IT adoption, more opportunities exist for market transactions than for transactions conducted in a business hierarchy (Malone, Yates, & Benjamin, 1987). These Information Technologies reduce transaction costs, bring customers and producers together, and promote online environments characterized by the elimination of traditional intermediaries. Accordingly, Benjamin & Wigand (1995) proposed that electronic commerce (EC) leads to the elimination of traditional players from the value chain, with direct buyer-seller interaction. For example, newspaper companies can provide news to the consumer using Internet without the newsagent participation. This phenomenon is called disintermediation, *displacement or* elimination of market intermediaries, enabling direct trade with buyers and consumers without agents (Wigand, 1997, p. 4).

In contrast, online environments also offer intermediation opportunities for new players that connect buyers and sellers and enable price searches (Bakos, 1991, 1997). These new types of intermediaries are called cybermediaries (Sarkar, Butler, & Steinfield, 1995). They are associated with new business opportunities related with the development of various intermediation functions on the Internet. In addition, they provide a range of services such as price fixing; transaction processing and coordination; stock management; quality guarantees; and monitoring (Spulber, 1996). This phenomenon is called re-intermediation or the creation of new intermediaries or market makers (Berthon, Ewing, Pitt, & Naudé, 2003, p. 555). Consequently, it is possible to describe the intermediation/disintermediation/re-intermediation cycle (Chircu & Kauffman, 1999), or web-enabled disintermediation and web-enabled re-intermediation in the web technology case (Berthon et al., 2003). See Table 1 for a summary of relevant disintermediation/re-intermediation debate research.

Table 1. Relevant disintermediation/re-intermediation debate research

Malone et al. (1987)	Electronic markets hypothesis (EMH): More opportunities for market transactions than for transactions conducted in a business hierarchy emerge with IT present. IT reduces transaction costs, promoting electronic markets characterized by elimination of traditional middlemen.
Bakos (1991, 1997)	Electronic markets offer intermediation opportunities for players that can help connect buyers and suppliers and enable price searches. Electronic markets offer lower search costs, thus enabling direct transactions and elimination of traditional middlemen.
Sarkar et al. (1995)	New types of intermediaries, called *cybermediaries*, emerge in electronic markets. Re-intermediation is associated with new business opportunities related with the development of various intermediation functions in cyberspace.
Benjamin and Wigand (1995)	EC leads to the elimination of traditional players from the value chain with direct buyer-supplier interaction.
Spulber (1996)	Intermediaries provide a range of services, including setting prices, transaction processing and coordination, inventory management, immediacy, quality guarantees and monitoring.
Bailey and Bakos (1997)	New roles for intermediaries in electronic markets include trust provision, aggregation, one-stop shopping, information exchange facilitators, and information filtering brokers. Filtering intermediaries can reduce information overload.
Bakos (1998)	Emerging electronic markets require intermediaries for matching, facilitation of transaction, and trust provision.
Prahalad (1998)	EC technologies require new intermediaries that provide expertise. EC technologies enable direct connections between buyers and suppliers.
Chircu and Kauffman (1999)	Technology providers have the ability to intermediate electronic markets by copying the roles of traditional intermediaries. Internet intermediaries replace the functions performed by established intermediaries, leading to their disintermediation. Internet intermediaries cannot sustain their first-mover competitive advantage. Established intermediaries have the ability to build their own EC systems and re-intermediate.
Hagel III and Singer (1999)	EC enables the emergence of profiling and information intermediaries, or *infomediaries*, who act as trusted agents on the behalf of clients.
Anderson and Anderson (2002)	The Internet was supposed to bring manufacturers into direct contact with end customers in a ruthlessly efficient world that left no room for middlemen. But the opposite is occurring. The downstream supply chain is becoming more complex and fragmented, creating opportunity for a new breed of uniquely positioned problem solvers.

The functions of these intermediaries have been amply studied in the literature (Adelaar, 2000; Anderson & Anderson, 2002; Bailey & Bakos, 1997; Bakos, 1997, 1998; Grover & Teng, 2001; Sarkar et al., 1995, 1998; Schmitz, 2000) and they can be summarized as follows: (1) Intermediaries reduce the number of buyer-seller required interactions, aggregating supply and demand, to obtain information from each other diminishing information asymmetries between them. Thereby, transaction costs are reduced; (2) Intermediaries reduce the cost of obtaining and processing existing information; (3) Intermediaries permit the interchange of information and coordination of transaction processes between market participants (buyers and sellers). By doing this, they reduce the associated information searching costs; (4) Intermediaries provide the technical, legal, and regulatory mechanisms that are required by market agents to conduct their transactions; and (5) Intermediaries provide mechanisms to ensure mutual trust and also the security of the transaction. They minimize agent opportunism and generate reputation and confidence in the channel, institution-based trust.

According to the roles or functions described previously, different types of intermediaries can be found (Rappa, 2004a, 2004b): marketplace exchange, buy/sell fulfillment, demand collection system, auction broker, transaction broker, distributor, search agent, virtual marketplace, and infomediaries, such as advertising networks, audience measurement services, incentive marketing, metamediaries, web aggregators, and syndicators.

In this chapter, we are interested in describing the Gb2bEM, as virtual marketplaces that provide institutional-based trust.

Global B2B Electronic Markets

The Gb2bEM can be defined as an online service to the benefit of buyers and sellers, offered by a neutral and independent market service provider against a fee (Perogianni, 2003), or as a place on the Internet, where many business buyers and suppliers meet, trade, and collaborate (Skjøtt-Larse, Kotzab, & Grieger, 2003). These intermediaries can be analyzed either from the Internet business models or from the inter-organizational trust perspective. In the first case, their characteristics as an Internet business model have been studied (Amit & Zott, 2000, among others), in particular its structure, content, and governance (Ordanini, Micelli, & Di Maria, 2004). In the second case, the trust building mechanisms used by Gb2bEM have been analyzed (Ba & Pavlou, 2002; Bolton, Katok, & Ockenfels, 2003; Dellarocas, 2003; Pavlou, 2002; Pavlou & Gefen, 2004) (see Figure 1).

Figure 1. Theoretical framework

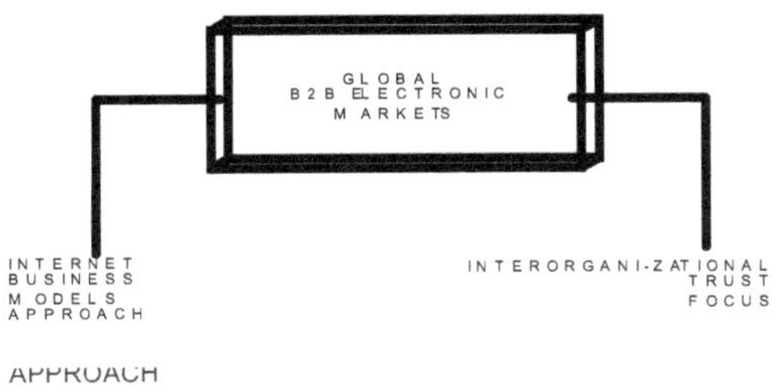

Internet Business Models Approach

From the Internet business models perspective, Gb2bEM are intermediaries, also known as e-marketplaces, market makers, e-hubs, or virtual vertical markets. These markets mainly arise from the evolution of B2B EC starting from EDI systems (electronic data interchange systems, originally they were one to one, costly and not scaleable), and changing towards current systems where many buyers and sellers can interconnect using Internet-based systems (Warkentin, Sugumaran, & Bapna, 2001). Nowadays, they become real online business communities that fulfil the needs of many companies, and facilitate the creation of value added among them (Santema & Van de Rijt, 2002).

Analyzing the Gb2bEM business model, three layers can be identified (Amit & Zott, 2001): content, structure, and governance.

The first layer, content, refers first, to the type of buyer or seller that operates on the Gb2bEM; they could be small and medium enterprises (SMEs) or big businesses (Ordanini et al., 2004) and second, to the business nature of Gb2bEM. They are the meeting point for collaboration (e-hubs collaboration) (Kalakota & Robinson, 2001) or platforms on the Internet where all the agents in the supply chain of an industry (sellers, manufacturers, importers, exporters, distributors, intermediaries, etc.) interchange information, carry out transactions, and collaborate. E-hubs are defined as intermediaries based on the Internet, concentrating on one industry or on one specific business process, which combine supply and demand and reduce transaction costs, and could be specific industry oriented (vertical) or business function oriented (horizontal) (Kaplan & Sawhney, 2000).

The layer governance refers to the ownership control and it is possible to identify two types of Gb2bEM (Ordanini et al., 2004; Riggins & Mitra, 2001; SkjØtt-Larse et al., 2003): neutral hubs (neutral pure intermediaries financed by financial institutions or others) and consortia (investors are the biggest participants in the industry). Two types of industrial consortia can be considered (Kalakota & Robinson, 2001): (1) consortia of buyers where a group of large companies combine their purchasing power in order to achieve price reductions; and (2) consortia of sellers that arise where there are few companies that control the market. They have the objective of avoiding price reductions and supplying more effective and efficient before-sale and after-sale support.

The third layer, structure, is defined by the differentiation through the developed revenue model (transaction or subscription) and the offered services to the market participants (catalogues or supply chain management services) (Berthon et al., 2003; Ordanini et al., 2004).

A description of some Gb2bEM, following the proposed three layers is on Table 2.

Table 2. Business model description (based on http://www.emarketservices.com/)

Gb2bEM	Content	Governance	Structure
Bizipoint	Manufacturers, importers, exporters, agents, traders. Electronics and Electrical industry	Omnitech Group Ltd.	*Revenue model*: hosting fee, the majority of the services are free although small fees are charged for value-added services. *Service portfolio*: Credit clearing, Industry directory, Industry news, Industry statistics, News/news letters, Procurement and fulfilment services.
Drug Discovery Online	Professionals in the drug development industry. Healthcare and Pharmaceutical industry	VertMarkets	*Revenue model*: Different fees for different services. *Service portfolio*: Bookstore, Collaboration tools, Consulting Services, Credit clearing, Discussion forum, Events calendar, Financing, Industry directory, Industry news, Job forum, Logistics, News/news letters, Trade publications and various software, Conversion tools, Market research reports and services to build an intranet.
Epapertrade.com	Professional paper buyers and sellers, such as producers, traders, paper merchants, printers or end users Paper and Pulp industry	Privately held	*Revenue model*: Transaction fee, Pre-agreed commission to the seller from finalized transactions. *Service portfolio*: Discussion forum, Logistics, Freight quotation, Payment arrangements and Advertisement.
Intercontinental Exchange	Financial institutions, power companies, energy producers and traders. Metal and Mining industry	BP Amoco, Deutsche Bank AG, The Goldman Sachs Group, Inc., Morgan Stanley Dean Witter, Royal Dutch/Shell Group, SG Investment Banking, and the Totalfina Elf Group.	*Revenue model*: Transaction fee. *Service portfolio*: Credit clearing, Events calendar, Industry news, News/news letters, Credit and risk management system, Order management system, Price indices.
ChemConnect	Worldwide chemicals, plastics, and industrial gas buyers and sellers. Plastics and Rubber industry	Privately held with major companies, such as Goldman Sachs, Institutional Venture Partners (IVP), BASF, Abbott Laboratories, as investors.	*Revenue model*: Membership fee, Transaction fee, Different fees for different services *Service portfolio*: Collaboration tools, Demotour, Events calendar, Industry news, Job forum, Logistics, News/news letters, Supply chain management, Payment and transaction monitoring and tracking, Financial Settlement.
CNTEX - China Textile Network	Sellers: Chinese textile and garment trade businesses and correlative manufacturers. Buyers: International companies interested in these products. Textiles & Leather industry	China Textile Network Co., Ltd (CNTEX) belongs to China National Textile Industry Council and China Textile Information Center	*Revenue model*: Membership fee, Three types of membership is available: Free, General, and Senior. Different annual fees for different services *Service portfolio*: Consulting Services, Events calendar, Industry directory, Industry news, Industry statistics, Logistics, News/news letters, Market prices, Credit standing evaluation, Expert forum, Investment guide, Translation services.
Ariba	Diverse industries such as consumer products, energy, financial services, high technology, manufacturing, pharmaceutical, public sector, higher education, telecommunications, and transportation are covered.	Privately held	*Revenue model*: Different fees for different services. Suppliers do not have to pay a fee *Service portfolio*: Collaboration tools, Consulting Services, Escrow, Events calendar, Industry news, Job forum, News/news letters, Private eMarket solution, Technical support, education, white papers, survey.

Table 2. continued

Gb2bEM	Content	Governance	Structure
B2GMARKET	Buyers: state, county and city governments of the U.S. and Canada, major international organizations such as the World Bank, ADB, UN, etc. Sellers: Companies globally searching for tender notices, government contracts, trade and project opportunities	Lufex, Inc.	*Revenue model*: Membership fee can be paid monthly or annually. A free trial period is available *Service portfolio*: Industry statistics, E-mail notification of new opportunities
Berlinkauf	Wide approach that includes multiple industries, such as automobile and vehicles, food and beverages, medical and pharmaceuticals, and minerals.	Not available	*Revenue model*: Membership fee, Different fees for different services *Service portfolio*: Industry directory, Industry news, Industry statistics, Logistics.
click2procure	Buyers are mainly Siemens different departments and daughter companies Suppliers are manufacturers, service providers, distributors who can meet Siemens procurement demands.	Siemens AG; Dep.: Siemens Procurement and Logistic Services	*Revenue model*: Membership fee, Basic functions for suppliers obligatory and free of charge, advanced services (e.g., participation at electronic auctions) for a fixed amount per year. *Service portfolio*: Consulting Services, Industry statistics, Supply chain management, Information services, catalogue management services, technical services for processing of market transactions or supplier integration service.
Global Trade Networks	A wide variety of products and services from electrical parts, and minerals, to textiles and wood products.	World Trade Marketing Limited an unlisted public company, registered in the state of NSW Australia.	*Revenue model*: Membership fee, Fee depends on the package chosen *Service portfolio*: Demotour, News/news letters, Translantion service, Currency converter, Web site hosting, Web development, online marketing, online payment facility through National Australia Bank, company research.
GNX	All types of products and services for the retail sector ranging from perishable foods to clothing.	Sears, Carrefour, Oracle, Coles Myer Ltd., KarstadtQuelle AG, The Kroger Co., METRO AG, Pinault-Printemps-Redoute, J Sainsbury plc and Federated Department Stores	*Revenue model*: Hosting fee, Membership fee, Different fees for different services *Service portfolio*: Collaboration tools, Consulting Services, Demotour, Events calendar, Industry directory, Industry news, Insurance, Job forum, News/news letters, Order tracking, Project management, Supply chain management, Online training seminars.

Another important aspect of the Gb2bEM research consists in the determination of success factors. Ordanini et al. (2004) applied the Amit and Zott (2001) description of an e-business model, to the case of Italian cases of b2bEM. They have found the consortia (with financial institutions or big companies from the sector as investors) as the most successful. However, this research does not consider trust as an explicative variable in the success model.

Inter-Organizational Trust Focus

In general, the buyer's power is growing in the digital economy. In this sense, Urban (2003) considers that businesses that cannot develop trust will have competitive disadvantages in the future; in fact, trust is even more essential in global EC (Ratnasingham, 1998). Furthermore, it is difficult for partners, mainly buyers and sellers, to build trust in global environments without face-to-face interaction because of the gap related to prior familiarity with each other; prior shared past experiences; and expectation of common future (Walther, 1995). In global B2B EC business partners collaborate with others through third-party structures or institution-based trust mechanisms, Gb2bEM.

Nevertheless, how do the Gb2bEM contribute to generate trust between buyers and sellers? Some authors have identified the trust-building mechanisms used by Gb2bEM to manage this objective.

The studied mechanisms are (Ba & Pavlou, 2002; Bolton et al., 2003; Dellarocas, 2003; Pavlou, 2002; Pavlou & Gefen, 2004):

- *Monitoring*: The Gb2bEM control sellers and take part in the resolution of conflicts, to ensure that the products have the appropriate conditions demanded by the market and ensure that transactions are realized in the correct way. This mechanism promotes a correct behaviour by buyers. In some markets a sanction-based system exists that can involve the expelling of the seller from the market for certain behaviours.
- *Accreditation*: The Gb2bEM act as an independent authorities certifying to the buyers that sellers adhere to certain rules. In this sense, the sellers are selected before operating on the market.
- *Legal bonds*: The Gb2bEM sign agreements with the obligations of sellers and the transactions are based on formal contracts between buyers and sellers.
- *Online feedback or reputation systems*: The Gb2bEM provide historic information about the transactions done in the market, informs the buyers about the opportunistic behaviour of some sellers (if they occur), and facilitate, to the buyers details of bad experiences with some sellers. This mechanism is similar to the word-of-mouth communication of off-line markets but supported by virtual environments.
- *Cooperative Norms*: The Gb2bEM facilitate the creation of rules and regulations that permit a quick resolution of conflicts between buyers and sellers, and the market facilitates all the necessary information to the buyers relative to the sellers before conducting the transaction.

Regarding trust, the research done by Pavlou (2002) and Pavlou and Gefen (2004) about Gb2bEM demonstrate how these intermediaries can positively influence the inter-organizational trust generation and favour the success of transactions in the case of the online buyer-seller relationships. In particular, in the sellers' community they can promote satisfaction, reduce the perception of the transaction-associated risk, and favour continuous relations. By extension, the G2b2EM can configure trust-based organizational networks. From a practical point of view, these G2b2EM can improve the design of their web pages to establish a better commercial environment as a first step. In addition, later they can promote the inter-organizational collaboration among those businesses increasing the volume of transactions. Finally, they have found that the success of this business model, and its existence as intermediaries, will depend on their capacity for generating commerce/collaboration environments based on the perception by the buyers/sellers community that they are online trust-building institutions (reputation online community perception).

Trust Building on Electronic Markets: Mechanisms and Results

The Gb2bEM, as Internet intermediaries, can generate institution-based trust and contribute to develop business communities in a first step and inter-organizational collaboration networks in a second step. According to the literature review, this is possible due to the use of some trust-building mechanisms to design the electronic markets. This allows companies to have a reputation between the buyers/sellers community, depending on the bigger or lesser-perceived efficacy of these trust-building mechanisms to conduct transactions in trusted environments (neutrals Gb2bEMs).

However, these Gb2bEMs can intermediate in a network of captive buyers and sellers (consortia of buyers or consortia of sellers). In this case, the trust-building mechanisms are not as necessary as they are in the prior case, because they are closed networks with organizations that collaborate in a virtual environment. With this, a research question for the future arises: Are the Gb2bEMs that use online trust-building mechanisms more successful?

According to the generated reputation between buyers and sellers, and to the characteristics of the business model's layers, a relations' model to explain the success of Gb2bEM is proposed (see Figure 2). In this model, the main elements and relations related to trust building on electronic markets are considered.

In order to measure the Gb2bEM success a latent variable is suggested. This Gb2bEM success will depend on the market incomes, number of transactions

Figure 2. Research model

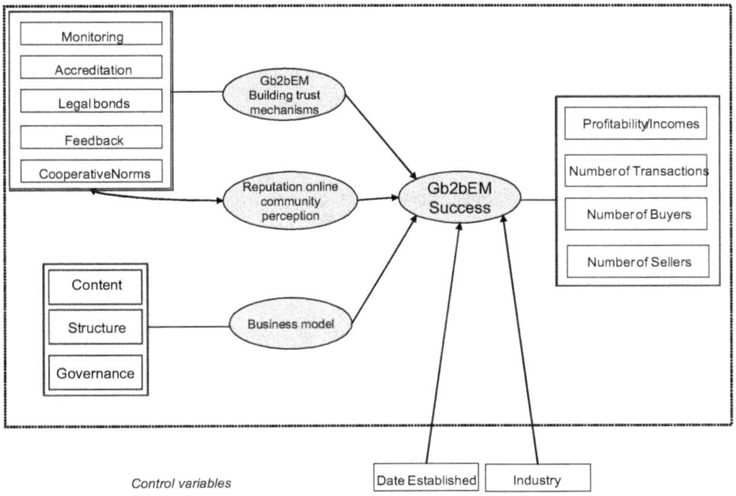

done in the market in a predetermined period, and the number of buyers and sellers who are members of the market. Lastly, as control variables, the starting date of the electronic market and its particular industry are proposed.

The objective in future investigations is to test this theoretical model (i.e., partial least squares, LISREL, etc.) and the realization of an empirical study, similar to Zaheer, McEvily, and Perrone (1998). There should be two types of respondents: the managers of the Gb2b2EM (for getting information about the trust-building mechanisms used in the market and the characteristics of their business model) and buyers and sellers of those markets (for collecting information about the perceived reputation by the agents according to the implemented trust-building mechanisms).

With this, the relation between the intermediation function developed by these companies as generators of institution-based trust and catalyst of international alliances and the success of Gb2bEM could be identified. In addition, other success factors related to the business model designed or apart from the proposed research model could arise.

Final Remarks

There is a debate about the disintermediation/re-intermediation process on the Internet. The fact is that new intermediaries came up with functions such as aggregating supply and demand; reducing the information costs; providing technical, legal, and regulatory mechanisms; and developing institution-based trust.

The Gb2bEM is an online service to the benefit of buyers and sellers, offered by a neutral and independent market service provider against a fee. Content, governance, and structure, as layers, can be used to describe the business model of these electronic markets.

One important factor for the success of these electronic markets is trust, and different types of trust-building mechanisms have been identified: monitoring, accreditation, legal bonds, online feedback/reputation systems, and cooperative norms. A model to explain the effects of the different trust-building mechanisms in the reputation building and in the development of inter-organizational collaboration networks is necessary. Building trust mechanisms, the perceived reputation online, and the characteristics of the business model affect the success of the Gb2bEM, measured through profitability and the number of transactions, buyers, and sellers of the market. Finally, empirical data will be necessary to test and complete the model.

References

Adelaar, T. (2000). Electronic commerce and the implications for market structure. *Journal of Computer-mediated Communication, 5*(3). Retrieved March 14, 2001, from http://www.ascusc.org/jcmc/vol5/issue3/adelaar.htm

Amit, R., & Zott, C. (2001). Value creation in e-business. *Strategic Management Journal, 22,* 493-520.

Anderson, P., & Anderson, E. (2002). The new e-commerce intermediaries. *MIT Sloan Management Review, 43*(4), 53-62.

Ba, S., & Pavlou, P. A. (2002). Evidence of the effect of trust building technology in electronic markets: Price premiums and buyer behaviour. *MIS Quarterly, 26*(3), 243-268.

Bailey, J., & Bakos, Y. (1997). An exploratory study of the emerging role of electronic intermediaries. *International Journal of Electronic Commerce, 1*(3), 7-20.

Bakos, J. Y. (1991, September). A strategic analysis of electronic marketplaces. *MIS Quarterly, 15*(3), 295-310.

Bakos, J. Y. (1997). Reducing buyer search costs: Implications for electronic marketplaces. *Management Science, 43*(12), 1676-1692.

Bakos, J. Y. (1998). The emerging role of electronic marketplaces on the Internet. *Communications of the ACM, 41*(8), 35-42.

Benjamin, R., & Wigand, R. (1995). Electronic markets and virtual value chains on the information superhighway. *Sloan Management Review, 36*(2), 62-72.

Benjamin, R., & Wigand, R. (1997). Electronic commerce: Effects on electronic markets. *Journal of Computer-mediated Communication, 1*(3). Retrieved January 5, 1998, from http://www.ascusc.org/jcmc/

Berthon, P., Ewing, M., Pitt, L., & Naudé, P. (2003). Understanding B2B and the web: The acceleration of coordination and motivation. *Industrial Marketing Management, 32,* 553-561.

Bolton, G. E., Katok, E., & Ockenfels, A. (2003, July). *Bridging the trust gap in electronic markets. A strategic framework for empirical study.* Working Paper. University Park, PA: Penn State University.

Buhalis, D. (2004). eAirlines: Strategic and tactical use of ICTs in the airline industry. *Information & Management, 41,* 805-825.

Chircu, A. M., & Kauffman, R. J. (1999). Strategies for Internet middlemen in the intermediation/disintermediation/reintermediation cycle. *Electronic Markets—The International Journal of Electronic Commerce and Business Media, 9*(2), 109-117.

Dellarocas, C. H. (2003). The digitalization of word-of-mouth: Promise and challenges of online feedback mechanisms. *Management Science, 49*(10), 1407-1424.

Grover, V., & Teng, J. T. C. (2001). E-commerce and the information market. Association for Computing Machinery. *Communications of the ACM, 44*(4), 79-86.

Hagel, III, J., & Singer, M. (1999). *Net worth—shaping markets when customers make the rules.* Boston: Harvard Business School Press.

Kalakota, R., & Robinson, M. (2001). *E-business. Roadmap for success.* Boston: Addison-Wesley.

Kaplan, S., & Sawhney, M. (2000, May-June). E-hubs: The new B2B marketplaces. *Harvard Business Review,* 97-103.

Malone, T. W., Yates, J., & Benjamin, R. I. (1987). Electronic markets and electronic hierarchies. *Communications of the ACM, 30,* 484-496.

Ordanini, A., Micelli, S., & Di Maria, E. (2004). Failure and success of B-to-B exchange business models: A contingent analysis of their performance. *European Management Journal, 22*(3), 281-289.

Pavlou, P. A. (2002). Institution-based trust in interorganizational exchange relationships: The role of online B2B marketplaces on trust formation. *Journal of Strategic Information Systems, 11,* 215-243.

Pavlou, P. A., & Gefen, D. (2004). Building effective online marketplaces with institution-based trust. *Information Systems Research, 15*(1), 37-59.

Perogianni, M. (2003). *B2B Internet trading platforms: Opportunities and barriers for SMEs. A first assessment.* Enterprise Papers, No. 13. Enterprise Directorate-General. European Commission. Belgium.

Porter, M. (2001, March). Strategy and the Internet. *Harvard Business Review,* 63-78.

Prahalad, C. K. (1998). Managing discontinuities: The emerging challenges. *Research Technology Management, 41*(3), 14-22.

Rappa, M. (2004a). *Business models on the web.* Retrieved January 20, 2004, from http://digitalenterprise.org/models/models.html

Rappa, M. (2004b). The utility business model and the future of computing services. *IBM Systems Journal, 43*(1), 32-42.

Ratnasingham, P. (1998). The importance of trust in electronic commerce. Internet research. *Electronic Networking Applications and Policy, 8*(4), 313-321.

Riggins, F. J., & Mitra, S. (2001). *A framework for developing e-business metrics through functionality interaction.* Working Paper. Atlanta: Georgia Tech.

Santema, S., & Van De Rijt, J. (2002). *The valuation of resources: Looking through the eyes of the customer.* Post-Conference Publication. Conference Competence 2002. Lausanne, Switzerland.

Sarkar, M. B., Butler, B., & Steinfield, C. (1995). Intermediaries and cybermediaries: A continuing role for mediating players in the electronic marketplace. *Journal of Computer Mediated Communication, 1*(3). Retrieved February 18, 1996, from http://www.usc.edu/dept/annenberg/vol1/issue3/sarkar.html

Sarkar, M. B., Butler, B., & Steinfield, C. (1998). Cybermediaries in electronic marketspace: Toward theory building. *Journal of Business Research, 41,* 215-221.

Schmitz, S. W. (2000). The effects of electronic commerce on the structure of intermediation. *Journal of Computer-Mediated Communication, 5*(3).

Retrieved March 14, 2001, from http://www.ascusc.org/jcmc/vol5/issue3/schmitz.htm

Skjøtt-Larse, T., Kotzab, H., & Grieger, M. (2003). Electronic marketplaces and supply chain relationships. *Industrial Marketing Management, 32,* 199-210.

Spulber, D. F. (1996). Market microstructure and intermediation. *Journal of Economic Perspectives, 10*(3), 135-152.

Urban, G. L. (2003, March). *The trust imperative.* Working Paper 4302-03. Cambridge, MA: MIT Sloan School of Management.

Walther, J. B. (1995). Relational aspects of computer-mediated communication: Experimental observations over time. *Organization Science, 6*(2), 186-203.

Warkentin, M., Sugumaran, V., & Bapna, R. (2001). E-knowledge networks for inter-organisational collaborative e-business. *Logistics Information Management, 14*(1/2), 149-162.

Wigand, R. (1997). Electronic commerce: Definition, theory, and context. *Information Society, 13,* 1-16.

Zaheer, A., McEvily, B., & Perrone, V. (1998). Does trust matter? Exploring the effects of interorganizational and interpersonal trust on performance. *Organization Science, 9*(2), 141-159.

Chapter IV

The Role of Shared Cognition and CSCW in Strategic Alliances

Michael A. Rosen, University of Central Florida, USA

Brooke C. Kelly, University of Central Florida, USA

Eduardo Salas, University of Central Florida, USA

Abstract

Information asymmetry is omnipresent within strategic alliances. This diversity in knowledge structure and content between team members comprising the alliance is complicated by the increasing trend of physical distribution of team members in international strategic alliances. While technical solutions to distance work abound, a strong and theoretically grounded approach to how best to implement these solutions is lacking. However, research in the area of shared cognition holds promise as an appropriate solution. This chapter presents a three-part framework for organizing research in the areas of shared cognition and CSCW (Computer Supported Collaborative Work) so that strategic alliance managers will have a resource for implementing their technologically mediated teams.

Introduction

Consider, if you will, an international strategic alliance based on a composite of several case study investigations. This hypothetical alliance, comprised of two manufacturing corporations, one U.S.-based and one based in France, each engaging in cooperation as a means to gain access to the other organization's market-specific information and thereby open new markets for its own products. Within this alliance, a globally distributed team (comprising members from the Chicago and San Francisco offices of the U.S. firm as well as members from the Paris and Toulouse offices of the French firm) was tasked with developing a marketing strategy for the U.S. firm's product line in France.

To begin their work, the team manager scheduled a videoconference between the four offices so that the team could ensure that everyone was on the same page as far as tasking and responsibilities. The meeting went well as a general introduction between the teams and produced a rough course of action. However, several weeks later, the team scheduled a second videoconference to work on some specific details of the marketing plan; this time, the meeting was unproductive. The setup of the videoconference made it difficult for team members to establish an orderly discussion. Team members from various locations would speak over other members and the discussion was difficult for many to follow. The videoconference ended with no progress being made. Initially, the team manger was stunned at the general failure of the second videoconference, but in retrospect he realized that the nature of the videoconference format did not afford the type of communication necessary to deliberate points of a specific plan. In a second effort to accomplish the original goal of a well defined plan of action, the team manager decided to use a threaded discussion board on the team's Web site so that team members could post ideas relevant to particular aspects of the plan and respond to other's ideas in an orderly and deliberative manner. The change of collaboration mode yielded returns in a rapid movement toward completion of the goal of the team and, subsequently, of the alliance.

Scenarios such as this are increasingly commonplace. Rapid growth in the amount of inter-organizational collaboration has been a consistent trend in the knowledge-based economy. Organizations continue to focus on and grow their core expertise, choosing to leverage that expertise with other specialized organizations in the form of strategic alliances. This paradigm maximizes the efficiency by which knowledge is brought to bear on the marketplace. However, the process of inter-organizational knowledge access is impeded by asymmetry in knowledge structure and content between team members constituting the alliance. A central benefit of the strategic alliance—access to expertise—is also a significant obstacle to the efficiency, effectiveness, and ultimate success of the strategic alliance collaboration when dealing with non co-located teams.

The problem of information asymmetry and distribution of expertise is compounded by the physical distribution of team members. A strategic alliance is comprised of individuals in different organizations and, ever increasingly, of individuals in different countries. These alliances do the majority of their work by means of collaborative technology and often lack the opportunity to engage in face-to-face communication. Despite the advances in communication technology and premature proclamations of the end of distance as an issue in the electronic age; research suggests that culture, language, social cues, context, and a host of other variables contribute to the persistence of physical distance as a mediator of teamwork (Olson & Olson, 2000; Gutwin & Greenberg, 2004). Therefore, if distance does truly matter, by what means can its effects be ameliorated? The ever-increasing need of an answer to this question in the management of international strategic alliances serves as the impetus for this chapter.

Therefore, in this chapter we explore how to design collaborative systems that will mitigate the effects of these two factors, distribution of space and expertise, in order to create effective and efficient teams and contribute to successful alliances. To this end, we present a conceptual framework for developing a method of implementing groupware systems based on the construct of shared cognition. Shared cognition is a term used to describe how "processes at the intraindividual level are dependent on and interact with process at the interindividual level" (Salas & Fiore, 2004, p. 138). Shared cognition helps to explain how teams are able to deal with dynamic, challenging tasks, and environments in an effective manner (Cannon-Bowers, Salas, & Converse, 1993). Team performance is a complex and multi-faceted concept, but for the parsimony, we offer a simple but powerful distillation of the literature. We draw from the team performance and CSCW research to present a framework for use in the task of managing strategic alliances. Specifically, in this chapter we seek to address the following key points:

- Clarify the connection between teams and strategic alliances;
- Introduce the concept of shared cognition as a means to overcoming performance deficits associated with information asymmetry in strategic alliances;
- Present an organizational shared cognition framework for conceptualizing factors contributing to a successful collaborative technology system; and
- Comment on specific needs for future research in order to develop empirically supported linkages in the framework.

Fundamentals of Shared Cognition and CSCW in Strategic Alliances

This section provides a brief introduction to the central concepts of the conceptual framework for implementing CSCW systems presented later in this chapter. Specifically, we will discuss the following constructs: teams, information asymmetry, shared cognition, shared mental models, transactive memory, and the current state of CSCW. We show conceptually that information asymmetry is always present in strategic alliances and serves both as the reason for the alliance and the most fundamental obstacle to its success. With the aim of using team research to address the issue of distributed expertise within strategic alliances, we illustrate that alliances are composed of teams and that team research findings and constructs are generalizable to alliances. Information asymmetry and analogous constructs serve as an active area of research in team science and this research remains a relatively untapped resource in addressing the fundamental problem of information asymmetry in strategic alliances. Lastly, this section reviews the fundamental dimensions of CSCW systems.

Strategic Alliances, Information Asymmetry, and Teams

The manner by which partnering organizations manage and develop their knowledge has been shown to be an integral component of the success or failure of that alliance (Larsson, Bengtsson, Henriksson, & Sparks, 1998). A recent theoretical conception of strategic alliances grounded in a knowledge-resource perspective has identified access of information as the primary driver of alliance formation (Grant & Badden-Fuller, 2004). That is, alliances represent a method of replacing costly organizational learning and transfer of skill with knowledge and information access. The knowledge resources available to an organization are thereby maximized while the cost of transferring that knowledge is simultaneously minimized.

Mutual learning is expensive in time and other resources. However, a situation where access to knowledge does not require a convergent growth of skills is far more efficient. In the simplest of terms, the strategic alliance allows for the least costly and most efficient application of knowledge to the marketplace. One organization enters an alliance with another when that organization possesses procedural or declarative knowledge that, in conjunction with its own knowledge, can be transmuted into viable products. For instance, the French and U.S. firms discussed in the introduction both sought to access specific market information held by the other in order to create opportunities for their own products in new markets.

The strategic alliance has an integral relationship not only with information asymmetry but with teams as well. Central to this chapter is the concept that the team is the molecular level of the larger (molar) strategic alliance unit. Alliances are composed of teams who, by the nature of alliances, are distributed in expertise and likely distributed in space. For our purposes, a team is "two or more people who interact, dynamically, interdependently, and adaptively toward a common and valued goal/objective/mission, who have each been assigned specific roles or functions to perform, and who have a limited life-span of membership" (Salas, Dickinson, Converse, & Tannenbaum, 1992, p. 4). The contact points between organizations in strategic alliances consist of teams whose effectiveness and efficiency contribute to the overall success of the alliance; therefore, strategic alliances are dependent upon effective and efficient team processes. The globally distributed team discussed in the introduction is a prime example of how the success of a strategic alliance is dependent (in part) upon the efficient functioning of its component teams. It is suitable then, to discuss team research within the context of strategic alliances and to apply theory and methods for overcoming incongruent mental models in teams, to the problem of information asymmetry in strategic alliances.

Strategic alliances between organizations share several characteristics of distributed teams. Distributed teams are defined as a "team or group whose members are mediated by time, distance, or technology" (Driskell, Radtke, & Salas, 2003, p. 297). Because these teams do not share physical space, distributed teams, like members of an alliance, must rely on some form of technology in order to aid communication (Priest, Stagl, Klein, & Salas, 2005). The effects of information asymmetry or incongruent knowledge structure on performance have been investigated extensively in team science. This research centers on an array of constructs captured under the umbrella of shared cognition. These constructs will be discussed in the following section and inform the framework presented later in the chapter.

Shared Cognition

A commonly used maneuver in basketball is the blind pass. When team members use this technique, the individual that possesses the ball is unable to see the intended receiver; instead he or she is estimating where the receiver will be. This estimation of a team member's location relies on all team members possessing an accurate and similar representation of the game. This shared representation of the game is a result of a team playing together for a length of time. So, what enables this fluid, implicit coordination?

Shared cognition enables teams to tacitly synchronize their behaviors, and has been shown to improve a team's performance (e.g., the blind pass leads to a basket). The term, shared cognition is used to describe the experience that occurs when members of a team share knowledge, ideas, and representations of various tasks or team elements. It is the degree to which team members have similar or compatible knowledge structures (Rentsch & Hall, 1994).

Shared cognition has been shown to be positively related to team process and team performance (Mathieu, Heffner, Goodwin, Salas, & Cannon-Bowers, 2000). Cannon-Bowers and Salas (2001) report that shared cognition leads to better task performance directly by increasing a team's precision, efficiency, quality of output, volume, and timeliness. Shared cognition also enhances team performance by improving the processes through which teams interact and by leading the teams to better motivational outcomes such as team cohesion, trust, morale, collective efficacy, and satisfaction (Salas & Fiore, 2004). There are several constructs that make up the larger subject of shared cognition including: shared mental models, transactive memory, and team situational awareness. These constructs represent different aspects of the global level shared cognition; they are parts of the whole. While related, these constructs represent distinct parts of the shared cognition concept.

Shared mental models are representations of the information that is shared by team members. This sharedness enables the team to effectively interact and process new information in a similar method. Shared mental models are knowledge structures held by team members that enable them to form accurate explanations and expectations for tasks and in turn, coordinate their actions and adapt their behavior to the demands of tasks and other team members (Cannon-Bowers et al., 1993). Shared mental models among team members are suggested to generate better decision making and improved coordination (Stout, Cannon-Bowers, & Salas, 1996). Shared mental models allow teams to draw inferences, better explain phenomena, make predictions, make decisions regarding action, and recognize relationships among aspects of the external environment (Cannon-Bowers et al., 1993; Mathieu et al., 2000).

Shared mental models are particularly beneficial under conditions where communication between team members is difficult (e.g., excessive workload, time pressure, distance). When teams are unable to engage in necessary strategizing, shared mental models become crucial to team functioning because members are able to predict information and resources required by their teammates (Mathieu et al., 2000). Thus, researchers argue that the presence of a shared mental model among team members accounts for the fluid, implicit coordination frequently observed in effective teams (Mohammed & Dumville, 2001).

Shared mental models are one feature of shared cognition. Another is transactive memory, a knowledge structure that enables team members to know what information is shared by the team and what information is held by each member as expertise (Kelly, Badum, Burke, & Salas, 2004). Austin (2003) contends that transactive memory systems provide group members access to a level of knowledge that no one could remember individually. The information shared in a transactive memory system can enhance a team's performance, particularly when the task is complex and requires a considerable contribution of knowledge from team members (Faraj & Sproull, 2000; Lewis, 2000). The specialization of knowledge an individual develops in a transactive memory system can reduce the cognitive load of individuals, provide access to an extended pool of expertise, and decrease redundancy of effort (Hollingshead, 1998). This particular aspect of shared cognition pertains to the shared and individual expertise that is located in a team. Another facet of shared cognition, team situational awareness, deals with the shared information that references things external to the team.

Team situational awareness is a shared perspective of the environment that provides integration and understanding of information. The concept of team situational awareness is especially important for the management of dynamic systems (Salas, Prince, Baker, & Shrestha, 1995). Teams in high-risk pressure situations (e.g., military, emergency and high hazard industries) need well-developed team situational awareness to successfully deal with a problem (Martin & Flin, 1997). Team situational awareness is not merely the sum of an individual's situational awareness or a complete group level idea of the situation, but an actively communicated and coordinated accomplishment between several members (Artman, 2000). Team situational awareness consists of both the situational awareness required of each team member and an overlap in situational awareness among team members for coordination.

Along with shared mental models and transactive memory, team situational awareness signifies evidence of the importance of shared cognition in teams. The research suggests that all concepts comprising shared cognition have a large impact on the performance and ultimate success of teams. The question then becomes, how do team members go about developing shared cognition? There are two documented methods in the literature: cross training and planning.

Cross training is a method that has been shown to contribute to the development of shared mental models between team mates (Marks, Sabella, Burke, & Zaccaro, 2002; Volpe, Cannon-Bowers, Salas, & Spector, 1996). Studies show it is possible to cross train using Distributed Mission Training (DMT) over distances to produce an effective composite training environment (Bell, 1999). DMT is the creation of a shared environment comprising real, virtual, and constructive systems that allow teams of individuals to train both individually and collectively (Carroll, 1999). DMT includes geographically dispersed individuals operating from different platforms at remote sites, who need to work together in a coordinated manner to achieve mission objectives (Ramesh & Andrews, 1999). In DMT information is dynamically shared and used among a group of individuals engaged in real-time mission-critical activities requiring coordination and communication (Carroll, 1999). DMT's components could constitute a training environment in any operational setting (Carroll, 1999).

Shared mental models can also be promoted through effective planning (Stout, Cannon-Bowers, Salas, & Milanovich, 1999), whereby the team can set goals; create an open environment; share information related to task requirements; clarify each team members' roles and responsibilities; discuss relevant environmental characteristics and constraints; prioritize tasks; determine what types of information all team members have access to and what types of information are held by only certain members; discuss their expectation; and pool unshared information. All of these activities help to promote a shared mental model through effective coordination.

Shared mental models, transactive memory, and team situational awareness are all aspects of the same construct—shared cognition. This construct is of central importance to the working of strategic alliances. The separation of strategic alliance team members in physical space and the gulf between them in expertise makes an understanding of shared cognition essential to the creation of effective and efficient alliances. By applying research in this domain, strategic alliance managers can harness the maximum yield of knowledge accessing between the information asymmetry for the lowest possible expenditure of training and mutual learning between the organizations.

However, there are factors that work against the success of an alliance team. Strategic alliance managers must contend with distribution in expertise (information asymmetry), which has been addressed in this section. The next section tackles the second issue central to information technology in strategic alliances—the physical distribution of team members in space.

Collaborative Technologies, Shared Cognition, and Teams

The physical proximity of teammates allows for faster development of congruent mental models (Bolstad & Endsley, 1999), but the time and cost of face-to-face interaction runs contrary to the efficiency aims and practical nature of strategic alliances. More recently, Social Network Distance (SND)—the amount of communication between team members—has been shown to be a predictor of mental model congruency in team members, independent of their physical proximity (Graham, Schneider, Bauer, Bessiere, & Gonzalez, 2004). To that end, a wide variety of information technology (IT) solutions have been developed to foster communication and collaboration. Over the past decade organizational conditions have become more favorable for the adoption of groupware and collaborative technologies. The products themselves have become more reliable and usable (Olson & Olson, 2003). However, the practice-based literature has not complemented the proliferation of collaboration options with methods of selecting modes of communication and collaboration that maximize shared cognition and consequently, performance.

At this point, it is necessary to more fully explore the actual dimensions by which a CSCW or groupware system can be implemented. "Groupware is software designed to run over a network in support of the activities of a group or organization" (Olson & Olson, 2003, p. 584). Users of groupware systems can differ in the time at which they access the system or the place from where they access the system. The two levels of time (synchronous or asynchronous) and space (distributed or co-located) form a general-purpose two-by-two matrix for classifying groupware systems. Within this matrix there are a multitude of modes of collaboration tools, such as e-mail, instant messaging, discussion boards, voice or videoconferencing, collaboratories, and shared knowledge management tools, to name but a few.

It is the role of the managers in strategic alliances to select the appropriate modes of collaboration, and it is the purpose of this chapter to work towards a framework for evaluating these options. Other research has applied constructs related to shared cognition (e.g., shared understanding and workspace awareness) to limited ranges of collaborative technology (e.g., shared visual spaces and collaboratories) (Arias, Eden, Fischer, Gorman, & Scharff, 2000; Gutwin & Greenberg, 2004). In the following section, we present a conceptual framework for organizing the research into a form useful to end users of CSCW systems in the field of strategic alliance management.

A Framework for Shared Cognition in CSCW Implementation

This chapter is an investigation in how best to solve problems of information asymmetry in strategic alliances by applying the construct of shared cognition to CSCW. In so doing, we find ourselves at an intersection of problem sets. That is, the ability to solve issues inherent in strategic alliances with CSCW and shared cognition relies logically first on the ability to address the issues within CSCW and shared cognition. This section first identifies a fundamental area of interest in CSCW research and application: an increase in the tailorability of systems. This issue is of particular importance to managers of strategic alliances because it suggests the future will require them to configure their own CSCW systems. Then, the issue of deciding how much and what type of shared cognition is appropriate for any given situation is addressed. We discuss why these problems are fundamentally linked and why work in these areas should be of concern to decision makers responsible for implementing information technology solutions for group work in strategic alliances. This section culminates with the presentation of a conceptual framework for organizing the research within these areas so that ultimately decision makers in strategic alliance management will have the requisite knowledge to implement CSCW systems that support the social processes conducive to effective and efficient teamwork.

Issues, Controversies, and Problems

In a recent analysis of historical trends within CSCW, Turoff, Hiltz, Fjermestad, Bieber, & Whitworth (2002) identify a pattern of increasing tailorability of CSCW systems. They extrapolate this trend and predict that more and more of the form and structure of a CSCW implementation will be placed in the hands of end users. Programmers and system designers will produce highly flexible groupware that enables the widest possible array of collaboration mode options. As a consequence, the group leaders or managers will be responsible for deciding how the group will interact via the technology.

In addition to how a collaborative system should be technically configured, when looking at CSCW and shared cognition from a pure efficiency perspective, the issue of how much shared cognition to support becomes important. How much is enough? How much is too much? Shared cognition is not *free*; it takes time and resources to develop. Training and supporting the amount and type of shared cognition that ultimately yields maximal team performance returns on the minimal investment remains an objective without a clear solution.

These issues are intertwined and comprise the what and how of implementing CSCW. First, the matter of type and quantity of shared cognition speak to what a system needs to achieve to create a successful collaboration. The increase in tailorability of CSCW systems implies a distribution of the procedural knowledge involved in designing groupware environments. More people with less training and experience in CSCW, will need to know how to select the processes essential to success, and in turn, what system components will support these team processes. This task of analyzing the team environment and creating the technical means of collaboration suited to that team will filter down to smaller and smaller units of management. The following section presents a framework that will assist in the successful completion of this task.

Solutions and Recommendations

The solution to this problem involves the development of a framework for organizing research in CSCW and shared cognition so that guidelines for system implementation can be culled from the literature and made available to end users in strategic alliance management. Figure 1 presents such a framework. The basic premise underlying this framework is that the managers of strategic alliance teams should be able to assess certain fundamental attributes of the team they are managing and then use that information to determine what aspects of shared cognition are most essential to support. Strategic alliance managers will then choose the CSCW system components that best facilitate these aspects of shared cognition and, consequently bolster team performance and effectiveness.

As defined previously, a team is comprised of team members (the composition of the team) who engage in tasks and work toward a collective goal. Hence, team composition, the task type, and team goals are the integral components in a discussion of team effectiveness. Therefore, as the fundamental aspects of a team, these three factors constitute the starting point of the framework and the input variables for strategic alliance managers. The input variables assist in making a decision about which aspects of shared cognition to support and consequently, which CSCW system components to implement. Each element of the framework is discussed.

Figure 1. Framework for shared cognition and CSCW.

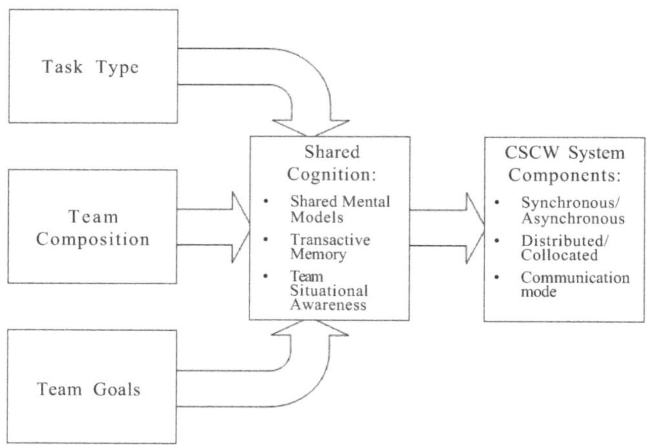

Task Type

The task type is the classification scheme for the short-term objectives that team members undertake while working toward the team goal. Small-group researchers have stressed that group processes and performance are strongly affected by the group task, and that understanding the nature of the task will help greatly in understanding team performance (Steiner, 1972). Tasks have been classified into schemes that organize tasks by things such as different classes of responses required for performance; hierarchical schemes (behavioral acts, behavioral requirements, and intrinsic task properties); and cycles of task events (inputs and outputs) (Dyer, 1984). Different types of intellectual small-group tasks (production, discussion, and problem solving) can account for up to 50% of the variance in group performance measures (Dyer, 1984). Shared mental models of a complex task are predictive of team performance. Therefore, as a defining characteristic of a team and considerable predictor of team performance, task type is an essential component of the framework.

Team Composition

In addition to information about the tasks being addressed by the team, strategic alliance managers will need to know basic information about the individuals in the team. Therefore, team composition is the second input variable to the framework and refers to the nature and attributes of group members: the team's knowledge, skills, and attitudes (KSAs). Team composition has been proposed to be an antecedent of the team's mental model (Edmondson, 1997). Heterogeneous (the mix of personalities, gender, attitudes, and background or experience) groups that initially performed poorly relatively to homogeneous groups later performed better than homogeneous groups on selected aspects of task performance (Guzzo & Dickinson, 1996). Heterogeneity is positively related to the creativity and the decision-making effectiveness of teams. Cooke, Kiekel, Salas, Stout, Bowers, and Cannon-Bowers (2003) have developed a set of measures for team knowledge and show that the information provided by these measures is predictive of future team performance. Hence, along with task type, the team composition is a critical element in predicting how the team will perform and how effectively the team will reach its goal.

Team Goal

The goal of the team is the purpose of the team, or what the team sets out to accomplish. Compared with the absence of goals (or the presence of ill-defined goals), specific, difficult goals for groups raise group performance on those dimensions reflecting the content of the goal (Guzzo & Dickinson, 1996). Therefore, the team goal represents an important descriptor of the team and an essential component in the framework. Task type, team composition, and team goal are fundamental attributes of a team and serve as a means of characterizing the strategic management team in this framework. The next two components of the framework, shared cognition and CSCW system components, flow from this characterization of the team and aim to facilitate team performance.

Shared Cognition

In this framework, shared cognition is an intermediate step between the input variables discussed above and the end result of a collaborative system that supports the needs of a specific strategic alliance team. The physical distribution of team members and the inherent information asymmetry between organizations involved in a strategic alliance, make the issue of supporting the aspects of shared cognition appropriate to task type, team composition, and team goal vital to the success of the team and ultimately the alliance. Although shared cognition is ultimately a single construct, the various aspects of shared cognition all have distinctive features and affect team processes and outcomes differently. Therefore, the central idea of the framework is that given certain information about the strategic alliance team, a designer can make predictions about the type of shared cognition most necessary for the successful completion of the team tasks and goals.

CSCW Components

Once the specifics of shared cognition have been prioritized, that is, which aspects are most essential to successful team functioning, the strategic alliance manager can then select the CSCW components that will facilitate the appropriate level and type of shared cognition. As discussed in the previous section, CSCW systems can vary on dimensions of synchronous/asynchronous communication, and collocated or distributed members. Each of the various modes of collaboration has an impact on the development of shared cognition. By developing linkages between the mode of collaboration and effects on shared cognition, CSCW designers will be able select designs to maximize team process and performance.

Summary of the Framework

The three input variables of task type, team composition, and team goal reflect the essential descriptive elements of a strategic alliance team. These descriptors can serve as the basis for the team assessment made by the strategic alliance manager, who then determines the aspects of shared cognition most essential to the particular team. Then the components of the CSCW system can be chosen that will facilitate development of the shared cognition relevant to task type, team composition, and team goal.

To illustrate how this model could be put to use, consider the example of the international alliance discussed in the introduction. Specifically, the team manager could have predicted the failure of the second videoconference simply by looking at the team goal, the type of task, and characteristics of the team's members. In this case, the type of task was the feature that made the videoconference a successful mode of collaboration for the first task, and unsuccessful for the second. The first task involved brainstorming and instilling a sense of common goals within the team and was suitable to the chosen mode of collaboration. The second task involved creating a detailed plan of action. Videoconferencing does not support the coherent communication of points of view on specific issues with larger numbers of people, nor does it allow the time necessary for team members to fully form their ideas or respond to the ideas of others. The asynchronous nature of the threaded discussion board allowed for better development of ideas and a better organization of the discussion. By accounting for task characteristics, the model presented in this chapter would have allowed the team manager to identify this issue and address it from the start.

Future Trends

As mentioned in the previous section, the tailorabilty of CSCW systems will continue to grow. Commercially available groupware will be malleable, and it will be the responsibility of the end user to customize the system to the needs of any one specific application; however, presently there is not a sufficient breadth of research or theory upon which to build hard and fast guidelines regarding implementation. The framework presented in the previous section is a step in that direction

Research is needed into the effects of mode of communication on the development of shared cognition as well as what types of tasks, goals, and team composition demand what type and amount of shared cognition. More is generally viewed as better, but from an efficiency perspective, shared cognition comes at a cost; it takes time to develop, therefore knowing the baseline of what is necessary for successful team performance is essential to implementing effective collaborative systems.

Conclusions

In international strategic alliances, the continuous trend of team members being distributed in space and expertise shows no sign of abating. Both the gulf in knowledge and physical distance between partnering organizations introduce separate sets of complexities. By emphasizing support for the development of specific aspects of shared cognition appropriate to the team task, goal, and composition, CSCW designers can help to realize the aim of strategic alliances: efficient transformation of knowledge into product for the marketplace.

References

Arias, E., Eden, H., Fischer, G., Gorman, A., & Scharff, E. (2000). Transcending the individual human mind—Creating shared understanding through collaborative design. *ACM Transactions on Computer-Human Interaction, 7*(1), 84-113.

Artman, H. (2000). Team situation assessment and information distribution. *Ergonomics, 4*(8), 1111-1128.

Austin, J. R. (2003). Transactive memory in organizational groups: The effects of content, consensus, specialization, and accuracy on group performance. *Journal of Applied Psychology, 88*(5), 866-878.

Bell, H. H. (1999). The effectiveness of distributed mission training. *Communications of the ACM, 42*(9), 73-78.

Bolstad, C. A., & Endsley, M. R. (1999). Shared mental models and shared displays: An empirical evaluation of team performance. In *Human Factors and Ergonomics Society 43rd Annual Meeting* (pp. 213-217). Santa Monica, CA: Human Factors and Ergonomics Society.

Cannon-Bowers, J. A., & Salas, E. (2001). Reflections on shared cognition. *Journal of Organizational Behavior, 22,* 195-202.

Cannon-Bowers, J. A., Salas, E., & Converse, S. (1993). Shared mental models in expert team decision making. In N. J. J. Castellan (Ed.), *Individual and group decision making* (pp. 221-246). Hillsdale, NJ: Erlbaum.

Carroll, L. A. (1999). Multimodal integrated team training. *Communications of the ACM, 42*(9), 69-71.

Cooke, N. J., Kiekel, P. A., Salas, E., Stout, R., Bowers, C., & Cannon-Bowers, J. (2003). Measuring team knowledge: A window to the cognitive underpinnings of team performance. *Group Dynamics: Theory, Research, and Practice, 7*(3), 179-199.

Driskell, J. E., Radtke, P. H., & Salas, E. (2003). Virtual teams: Effects of technological mediation on team performance. *Group Dynamics: Theory, Research, and Practice, 7*(4), 297-323.

Dyer, J. L. (1984). Team research and team training: A state of the art review. *Human Factors Review, 8*, 285-319.

Edmondson, A. C. (1997, April). *Group and organizational influences on team learning.* Dissertation Abstracts International: Section B. *The Sciences & Engineering, 57*(10-B), 6647.

Faraj, S. A., & Sproull, L. (2000). Coordinating expertise in software development teams. *Management Science, 46,* 1554-1568.

Graham, J., Schneider, M., Bauer, A., Bessiere, K., & Gonzalez, C. (2004, September 20-24). *Shared mental models in military command and control organizations: Effect of social network distance.* Paper presented at the Human Factors and Ergonomics Society 48th Annual Meeting, New Orleans, LA.

Grant, R. M., & Baden-Fuller, C. (2004). A knowledge accessing theory of strategic alliances. *Journal of Management Studies, 41*(1), 61-84.

Gutwin, C., & Greenberg, S. (2004). The importance of awareness for team cognition in distributed collaboration. In E. Salas & S. M. Fiore (Eds.), *Team cognition: Understanding the factors that drive process and performance* (pp. 177-201). Washington, DC: American Psychological Association.

Guzzo, R. A., & Dickinson, M. W. (1996). Teams in organizations: Recent research on performance and effectiveness. *Annual Review of Psychology, 47,* 307-338.

Hollingshead, A. B. (1998). Retrieval processes in transactive memory systems. *Journal of Personality and Social Psychology, 74*(3), 403-437.

Kelly, B. C., Badum, A., Burke, C. S., & Salas, E. (2004, April 1-4). *Shared cognition: Can we all get on the same page?* Paper presented at the Meeting of the Society of Industrial and Organizational Psychology, Los Angeles, CA.

Larsson, R., Bengtsson, L., Henriksson, K., & Sparks, J. (1998). The interorganizational learning dilemma: Collective knowledge development in strategic alliances. *Organization Science, 9*(3), 285-305.

Lewis, K. (2000, August 4-9). *Transactive memory and performance of management consulting teams: Examining construct and predictive validity of a new scale.* Paper presented at the Annual Meeting of the Academy of Management, Toronto, Canada.

Marks, M. A., Sabella, M. J., Burke, C. S., & Zaccaro, S. J. (2002). The impact of cross-training on team effectiveness. *Journal of Applied Psychology, 87*(1), 3-13.

Martin, L., & Flin, R. (1997). *Building fire commander situational assessment from team shared mental models.* Paper presented at the IEE Colloquium on Computer Mediated Complex Supervisory and Decision Making in Teams, Glasgow, Scotland.

Mathieu, J. E., Heffner, T. S., Goodwin, G. F., Salas, E., & Cannon-Bowers, J. (2000). The influence of shared mental models on team process and performance. *Journal of Applied Psychology, 85*(2), 273-283.

Mohammed, S., & Dumville, B. C. (2001). Team mental models in a team knowledge framework: Expanding theory and measure across disciplinary boundaries. *Journal of Organizational Behavior, 22*(2), 89-103.

Olson, G. M., & Olson, J. S. (2000). Distance Matters. *Human-Computer Interaction, 15*(2-3), 139-178.

Olson, G. M., & Olson, J. S. (2003). Groupware and computer-supported cooperative work. In J. A. Jacko & A. Sears (Eds.), *The human-computer interaction handbook: Fundamentals, evolving technologies and emerging applications* (pp. 583-595). Mahwah, NJ: Lawrence Erlbaum Associates.

Priest, H. A., Stagl, K. C., Klein, C., & Salas, E. (2005). Creating context for distributed teams via virtual teamwork. In C. A. Bowers, E. Salas, & F. Jentsch (Eds.), *Creating high tech teams* (pp. 185-212). Washington, DC: American Psychological Association.

Ramesh, R., & Andrews, D. H. (1999). Distributed mission training: Teams, virtual reality, and real-time networking. *Communications of the ACM, 42*(9), 65-67.

Rentsch, J. R., & Hall, R. J. (1994). Members of great teams think alike: A model of team effectiveness and schema similarity among team members. In M. M. Beyerlein & D. Johnson (Eds.), *Advances in interdisciplinary studies of work teams: Theories of self-managing work teams* (Vol. 1, pp. 223-261). Greenwich, CT: JAI Press.

Salas, E., Dickinson, T., Converse, S., & Tannenbaum, S. (1992). Toward and understanding of team performance and training. In R. Swezey & E. Salas (Eds.), *Teams: Their training and performance* (pp. 3-29). Norwood, NJ: Ablex Publishing.

Salas, E., & Fiore, S. M. (2004). Why team cognition? An overview. In E. Salas & S. M. Fiore (Eds.), *Team cognition* (pp. 3-8). Washington, DC: American Psychological Association.

Salas, E., Prince, C., Baker, D. P., & Shrestha, L. (1995). Situation awareness in team performance: Implications for measurement and training. *Human Factors, 37*(1), 123-136.

Steiner, I. D. (1972). *Group processes and productivity*. New York: Academic Press.

Stout, R. J., Cannon-Bowers, J. A., & Salas, E. (1996). The role of shared mental models in developing team situational awareness: Implications for training. *Training Research Journal,* 2, 85-116.

Stout, R. J., Cannon-Bowers, J. A., Salas, E., & Milanovich, D. M. (1999). Planning, shared mental models, and coordinated performance: An empirical link is established. *Human Factors, 41*(1), 61-71.

Turoff, M., Hiltz, S. R., Fjermestad, J., Bieber, M., & Whitworth, B. (2002). Computer-mediated communications for group support: Past and future. In J. M. Carroll (Ed.), *Human-computer interaction in the new millennium* (pp. 279-302). New York: ACM Press.

Volpe, C. E., Cannon-Bowers, J. A., Salas, E., & Spector, P. E. (1996). The impact of cross-training on team functioning: An empirical investigation. *Human Factors, 38*(1), 87-100.

Chapter V

Sharing Knowledge in Strategic Alliances to Build Collaborative Advantage

Fátima Guadamillas Gómez,
University of Castilla-La Mancha, Spain

Mario Javier Donate Manzanares,
University of Castilla-La Mancha, Spain

Jesús David Sánchez de Pablo González del Campo,
University of Castille-La Mancha, Spain

Abstract

This chapter studies the transmission and sharing of knowledge in strategic alliances. First of all, the authors explain the existence of alliances from different theoretical perspectives. They argue that resource and knowledge-based views of the firm offer an adequate approach for the understanding of social and strategic aspects involved in the achieving of a collaborative advantage. Secondly, they examine information technology (IT) tools that are critical for enabling and supporting information and knowledge-sharing processes among partners. Also, they emphasize other important problems that make the effective development of collaborative activities and knowledge sharing in strategic alliances difficult: distrust between partners and cultural barriers in relation to collaboration; the excessive technological and knowledge diversity or similarity among partners that makes learning more difficult; and the governance structure of the alliance. Finally, some practical solutions are proposed: the correct use of IT tools, some organizational measures, human resource practices, and the limitation of the scope of the agreement.

Introduction

In recent years, uncertainty, volatility, and risk are some factors that better define the current competitive environment. Firms in most industries carry out restructuring processes in order to face this situation and alliances become a key tool for competitiveness of firms in the knowledge society.

Alliances can be a source for firms to acquire and improve their knowledge-based capabilities in the current knowledge-intensive environment (Oxley & Sampson, 2004). They can also be a mechanism for firms to build a competitive advantage and to gain or maintain leads over their rivals because of their proven access to economies of scope, scale, complementary capabilities, and knowledge.

In this work, we study the transmission and sharing of knowledge in alliances from the knowledge-based view of the firm and the role of information technologies in this process. The paper is divided into four parts. First of all, we explain the importance for firms of sharing knowledge in alliances to build collaborative advantage and to obtain higher returns on their R&D investment. Secondly, we look at the importance of IT in this process. Third, we try to study the factors that complicate the effective transmission and sharing of knowledge. Finally, some ideas are proposed with the purpose of trying to solve problems associated with knowledge transmission in strategic alliances.

Building Collaborative Advantages in Strategic Alliances

When firms undertake alliances or cooperation agreements they have to face some practical questions that basically can be grouped into two blocks (Navas

Table 1. Collaborative advantages in strategic alliances

Advantage: Access to complementary resources and capabilities and knowledge transfer
ACTIONS
* Access to more complex resources and capabilities that firms lack * Access to a greater variety of technology that is converging in an industry * Acquisition of new knowledge from the partners * Knowledge transfer
REFERENCES
Kogut (1988); Hamel (1991); Nohria and Garcia-Pont (1991); Inkpen and Beamish (1997); Nohria and Piskorski (1997); Parkhe (1998); Koka and Prescott (2002)
Advantage: Entrance into new markets or businesses
ACTIONS
* Collaboration with local partners to get access to new knowledge * Reduce uncertainty
REFERENCES
Kogut (1988); Hamel (1991)
Advantage: Increase competitive power
ACTIONS
* Increase negotiation power * Achieve commercial or technology standards
REFERENCES
Pfeffer and Salancik (1978); Kogut (1988); Gulati (1998)
Advantage: Cost reduction/ Efficiency increase
ACTIONS
* Economies of scope, scale, and learning * Coordination of activities and complementary capabilities * Reduce risk for sharing investments * Reduce transaction costs, achieving or keeping, simultaneously, a high flexibility
REFERENCES
Kogut (1988); Hagedoorn (1993)
Advantage: Management
ACTIONS
* Development of other strategic options that can not be reached in an isolated way such as diversification or internationalisation * Obtain synergies * Improve reputation, image or prestige
REFERENCES
Hamel (1991); Hagedoorn (1993); Grant (1996); Saxton (1997); Lane and Lubatkin (1998); Gulati (1998); Stuart (2000)

Table 1. continued

Advantage: Learning
ACTIONS
* Learning from the management of an alliance to achieve higher efficiency in future cooperation agreements * Learning from the characteristics of a market * Acquisition of tacit and technological knowledge and new abilities
REFERENCES
Hamel (1991); Hagedoorn (1993); Hagedoorn and Schakenraad (1994); Gulati (1998); Lane and Lubatkin (1998); Parkhe (1998); Stuart (2000)

& Guerras, 2002, p. 450): questions about the process of negotiation of the agreement—activities that can be carried out before the alliance begins to work in an effective way—and activities related to the management of the agreement. These two blocks of questions involve a group of actions that a potential partner has to undertake, such as to look for, select, and get in touch with other potential partners (according to the pursued objectives); set up the conditions of development of the agreement; control the process; and evaluate the objectives reached.

The way in which firms achieve cooperation agreements and manage those activities determines their advantages and disadvantages or risks. Owing to this, it is necessary to carry out an in-depth analysis of the generation of advantages in strategic alliances, specifically those derived from knowledge creation and transmission. The main advantages that justify participation in strategic alliances are shown (Table 1).

Lane and Lubatkin (1998) and Stuart (2000) affirm that the main objective of partners in an alliance is inter-organizational learning, as a consequence of the difficulty of solving internal environmental problems for each firm. Inter-organizational learning is based in the absorptive capacity, which is the ability of the firm to value, assimilate, and make use of new external knowledge (Cohen & Levinthal, 1990). Absorptive capacity depends on the knowledge bases, organizational structures, and dominant logics of firms that cooperate (Lane & Lubatkin, 1998). Moreover, firms with a greater absorptive capacity will invest more on R&D through alliances (Cohen & Levinthal, 1990). Know-how access is another main reason for firms in technology-intensive industries to carry out cooperation agreements, because it makes possible the quicker development of innovations and makes easier their market introduction in a short period of time, giving important advantages and encouraging learning (Stuart, 2000).

Stuart (2000) points out another important advantage: the improvement of reputation, which makes a more favourable competitive position possible. The better the reputation achieved by a firm, the greater access it has to the knowledge sources of other organizations and it becomes an "attractive" partner in terms of its cumulated experience in cooperation (Gulati, 1998, p. 296). But reputation is an intangible resource that is built slowly but can be quickly destroyed. Due to this, alliances contribute towards avoiding the development of an opportunistic behaviour (Gulati, Nohria, & Zaheer, 2000). This behaviour carries a high cost due to the damage that it can cause to the firm's reputation not only by damaging current agreements but also those that could be reached in the future (Gulati et al., 2000).

Table 2. Risks involved in strategic alliances

RISK	CONSEQUENCES	REFERENCES
Existence of organizational asymmetries shown in differences on beliefs, values, organizational routines, language, clients, and traditions.	Unequal delivery between partners	Kogut (1988); Inkpen and Beamish (1997); Rialp (1999); Stuart (2000)
	Uncertainty and opportunistic behaviour increases	
	Difficulties in communication and transference of knowledge	
	Cultural ambiguity	
Variability of the environment	Instability	Chung, Singh, & Lee (2000)
	Lack of will to give resources due to uncertainty.	Rothaermel and Deeds, (2004)
Power dynamic	Struggle to attain maximum learning	Hamel (1991)
Exclusive realization of internal analysis	Not to take advantage of environmental opportunities	Gulati (1998)
Differences in information transmission	Struggle to improve position in the alliance	Gulati (1998); Soh (2003)
Lack of flexibility in the management of the alliance	Not to take advantage of opportunities and cost increase	Gulati (1998)

Kogut (1988), Hamel (1991), and Inkpen and Beamish (1997) highlight the importance of other advantages such as the intention of sharing costs and risks with partners, along with access to complementary resources, new markets, and products. Companies opt for cooperation when they do not possess necessary resources and capabilities in order to develop the optimal strategy; these cannot be internally developed or externally acquired (Montoro, 2000; Rialp, 1999). Simultaneously, cooperation avoids duplication of expenses, giving scale economies and risk dispersion. Finally, if the development of agreements is done in an effective manner and the objective pursued is access to new markets, incomes can be increased by a considerable amount through access to new market niches, where the company has not previously operated, making easier the access to a larger portfolio of clients (Stuart, 2000).

However, the attainment of collaborative advantages presents some considerable risks. Table 2 shows risks of cooperation and their fundamental consequences in achieving a better understanding of strategic alliances.

These risks are mainly related to the existence of information asymmetries and differences in information transmission that can cause uncertainty and opportunism to increase (Gulati, 1998). Also, important differences arise in relation to beliefs, values, organizational routines, and languages, which make communication and knowledge transmission difficult (Stuart, 2000, p. 808). Other risks have to do with the power dynamic and the struggle to attain more learning than a competitor, not to take advantage of environmental opportunities and the lack of flexibility in the management of the alliance, which can cause a cost increase (Gulati, 1998). Risks that have a more direct effect on knowledge transmission and sharing—the main point of the chapter—will be analysed next.

After taking advantages and risks associated with cooperation agreements, firms have to evaluate whether establishing alliances is advisable or not. The company has more accurate knowledge of the possibilities of the alliance and can better plan its conception and design. This contributes to the effective creation of collaborative advantages.

The main advantages of cooperation agreements are analysed in following sections, focusing on those related to knowledge management originating from strategic alliances. At the same time, the organization has to make suitable decisions on the optimal way of participation in the alliance according to the advantages that cooperation can give and the objectives found: partners, scope, time, object of the agreement, and knowledge available to share. Also, partners have to look for tools that allow the reduction of risks that have to be faced in the development process of the agreement. One of the essential tools to accomplish the objective of the agreements and reduce the risks may be IT, the role of which in relation to cooperation agreements will be analysed in following sections.

Theoretical Perspectives about Strategic Alliances

Reasons that justify the establishment of strategic alliances basically come from three different perspectives: (1) economics, (2) strategic management, and (3) organizational theory (Menguzzato, 1995). According to the first perspective and mainly from the transaction cost theory, hierarchical governance structures (i.e., firms) are better than market contracting when opportunistic behaviours are probable, and therefore, transaction costs are high (Williamson, 1975). If this reasoning is applied to alliances, these will be useful in middling situations, that is to say, when transaction costs are not as high as to require a hierarchical control but not as low as to prefer an exchange base in the market (Gulati et al., 2000).

The organizational theory justifies the possibility that cooperation can be an alternative way to cover the firm's need to obtain and develop a group of necessary resources for its survival. This is an alternative way of obtaining resources to market acquisition and/or internal development (Navas & Guerras, 2002, p. 428), especially when reciprocity or complementarities in the availability of resources—knowledge—exists (Bowman & Faulkner, 1997).

Finally, from the viewpoint of strategic management, it is argued that alliances base their logic on the possibility of attaining a better competitive position when it is not possible to achieve it through any other means (Navas & Guerras, 2002, p. 428). From this perspective, we found the most important reason for the existence of strategic alliances, especially from the theoretical perspective of the resource-based view of the firm (Barney, 1991; Peteraf, 1993), the knowledge-based approach (Grant, 1996; Kogut & Zander, 1992; Nonaka & Takeuchi, 1995; Spender, 1996), and the social capital perspective (Dyer & Singh, 1998; Nahapiet & Ghoshal, 1998).

These perspectives give different—and sometimes stronger—points of view than the traditional explanation provided by the transaction cost theory. Explanations for the formation of alliances based on the transaction cost perspective, which emphasized the achievement of efficiency in transaction costs, are more suitable when there are routine situations between partners or the intention is to obtain a static efficiency and other strategic advantages, such as inter-organizational learning, legitimacy creation, or quick entrance into the market are put aside (Eisenhardt & Schoonhoven, 1996, p. 137). In view of this, Eisenhardt and Schoonhoven (1996, p. 136) remark that transaction costs theory is unable to capture strategic and social aspects that push the majority of firms into forming alliances, which previously explained theories based on strategic management are able to offer.

From the resource-based approach, the main objective of alliances is value creation through the development and exploitation of resources and capabilities. Sometimes, the only way for the firm to access necessary resources is through cooperation, because these resources are rooted in other firms (Chung et al., 2000).

Moreover, from the resource-based perspective, it is cited that firms which are able to develop and manage alliances (or the firm's network) better than competitors—which Kale, Singh, and Perlmutter (2000) address as the firm's alliance capability—can achieve a sustained competitive advantage unless competitors can imitate the activities—that is, alliance management—that allow the firm to create value (Ireland, Hitt, & Vaidyanath, 2002). Also, the greater dynamic within the environment, the smaller the possibility of the firm having all the necessary resources available, and in this respect, alliances give access to these resources and capabilities and they are a continuous source of learning as well (Ireland et al., 2002). So, from this perspective, an important source of creation of valuable and inimitable resources arises from the network of relationships established by the firm (Gulati et al., 2000, p. 207).

Relational rents can be understood as "supernormal profit jointly generated in an exchange relationship that can not be produced by either firm in isolation and can only be created through the joint idiosyncratic contributions of the specific alliance partners" (Dyer & Singh, 1998, p. 662). Management of strategic alliance*s* involves four aspects that directly affect the possibility of achievement of relational rent (Dyer & Singh, 1998): (1) investments in relation-specific assets; (2) management of knowledge exchange, including that which results in joint learning and routines; (3) adequate combining of complementary resources and capabilities that result in the joint creation of new products, services, or technologies; and (4) effective governing mechanisms of inter-firm alliance that result in a lower transaction cost than competitor cooperative arrangements.

As an extension of the resource-based perspective, the knowledge-based view of the firm regards organizational knowledge as an important factor in contributing to a firm's pursuit of competitive advantage (Foss, 1999; Grant, 1996; Liebeskind, 1996). Thus, knowledge can and must be managed in order to improve firm performance (Oliveira, 1999).[1]

The Role of Information Technologies in Strategic Alliances

ITs can play a critical role as a supportive tool in alliances and they can be said to help to explain the fast growth of networks in the last few years (Gulati et al.,

2000). The utility of IT can be considerable in the management of some of the more important tasks of the agreement, such as the transmission and storage of knowledge and the outcome of monitoring of activities.

Previously, the justification for the existence of strategic alliances has been established from the economics, strategic management, and organizational theory perspectives. The importance of IT tools in cooperation agreements is defended, based on the same theoretical grounds. In relation to the economic explanation—mainly the transaction cost theory—the main profit related to IT is the transaction cost reduction of the inter-organizational relationship as compared to what it would be in market terms. It is important to note that this is not to say that transaction cost in the market cannot be reduced (because of IT), but cooperation and the use of IT in alliances can reduce transaction cost more than the use of these in market relationships.

For example, Medina, Bruque, and Ruiz (2003) point out that IT tools can reduce transaction costs in inter-organizational networks, and they analyse their impact on coordination costs and transaction hazards, which are both components of transaction costs in durable inter-organizational relationships (Clemons & Row, 1992). In relation to this, information systems provide a higher capacity at low cost to supervise relationship performance. Also, strategic alliances reduce transaction hazards because of trust generated among partners. In this sense, IT tools allow the storage of information about partners and thus, diminish organizational asymmetries. In addition to this, IT tools reduce the loss of information resources in the network (Clemons & Row, 1992). At the same time, IT tools also contribute to leverage the value created in alliances through the joint use by partners of design, engineering or computer assistance tools in production (Arjonilla & Medina, 2000). In this case, logic comes from the synergy advantages obtained through the concentration of each partner in their core resources and capabilities (i.e., things that they know how to make better).

From a strategic and organizational point of view, we find that the use of IT tools have a greater potential outcome than the transaction cost perspective is able to explain, due to the fact that IT tools are a critical aspect of the organizational knowledge management activities (knowledge and information access, transfer, sharing, and storage). IT tools play a critical role in the management of the organizational knowledge owing to the fact that knowledge, under some conditions (i.e., codification), can be treated as an object that can be divided in modules (structured), gathered, and transferred (Sanchez & Mahoney, 1996; Zander & Kogut, 1995).

Thus, knowledge management and the advantages of IT application in alliances and networks are, firstly, related to the management of the firm's social capital as an intensive resource in knowledge and information.[2] Secondly, they are critical for the building, improvement, and use of valuable inter-firm specific-knowledge sharing routines that are difficult to imitate and to substitute (Dyer &

Singh, 1998). Due to the importance of both mechanisms, they will be analysed in more detail.

Social Capital Management and IT

The understanding of IT support in the generation and management of the firm's social capital implies, first of all, defining the concept. Thus, the firm's social capital can be roughly understood as the goodwill—referring to the sympathy, trust, and forgiveness offered us by friends and acquaintances—that is engendered by the fabric of social relationships and that can be mobilized to facilitate action (Adler & Kwon, 2002, pp. 17-18). Now that goodwill is the substance of social capital, its effects flow from the information, influence, and solidarity that such goodwill makes available, and their sources lie in the social structure within which the firm is located (Adler & Kwon, 2002, p. 18).[3]

However, researchers consider social capital as a multi-dimensional concept which covers, in a general way, three different areas (Lee, Lee, & Pennings, 2001; Nahapiet & Ghoshal, 1998): (1) structural, referring to network ties—information transmission channels—and network configuration—its density, connectiveness, and hierarchy—; (2) cognitive, which refers to the need of a common context for communication development, within which language and shared codes are included; and (3) relational, referring to a more emotional dimension of social capital such as trust that is generated in social relationships.

Both development—management of strategic alliances—and development of social capital have a mutual influence, and they are interrelated in an ongoing way, as different contributions from knowledge-based and social-capital perspectives show (Adler & Kwon, 2002; Hitt, Ireland, & Santoro, 2004). For example, value created in alliances, that is, improvement of economic performance; resources and capabilities building; learning improvement; and enlargement of the strategic network (Hitt et al., 2004) has an important influence on the firm's relationships network because it provides the trust of future partners, due to the reputation acquired; the improvement in channels of communication; and the development of a shared code among partners, which no doubt increases the possibility of accessing new partners and resources (Dyer & Singh, 1998; Gulati et al., 2000).

Moreover, social capital affects the management of new alliances, for example, in placing the firm in the best position to find an adequate partner or because of the experience to define the right scope of the agreement; technologies to be employed and developed; markets to be served; boundary configurations; governance structure to use; and so forth. (Hitt et al., 2004, p. 18). Also, beyond their mutual influence, social capital has a mediate position between the

Figure 1. Relationship between alliance development and management, social capital and value creation (Hitt, Ireland, & Santaro, 2004)

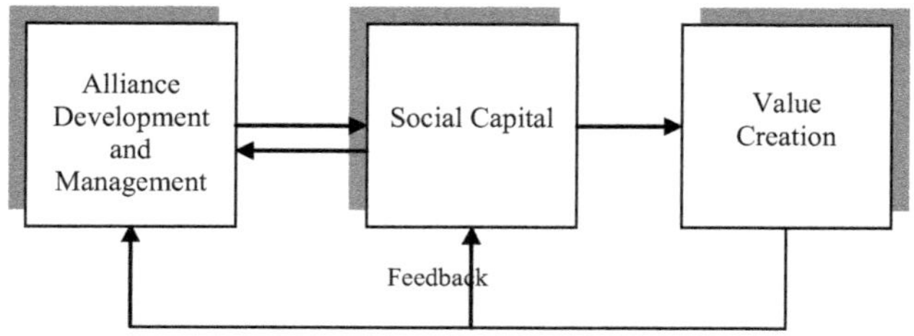

management of the alliances' capability and value created (referring to the management of the alliance). Moreover, a feedback dynamic exists among all of these; that is, that market power, financial wealth, and reputation gains obtained from cooperative arrangements through enhanced resources, new learning, and an enlarged strategic network improve the firm's capability and its social capital (Hitt et al., 2004), as shown in Figure 1.

The role of IT in the management of social capital is basically related to structural and cognitive dimensions. Thus, the network and information (knowledge) access through technological tools at a reduced cost, and the storage and structuring of knowledge are some aspects that impact on the quality of the connectiveness and communication channels and allow greater speed in access to resources and partners. On the other hand, the establishment of interfaces, in an endeavour for the development of a common code, supports the cognitive scope of social capital for enhancing the quality of interactions and knowledge transfer and sharing. For relational dimension, IT tools have a lower applicability due to the fact that it implies higher levels of an emotional component that reduces their potential use.

Building, Improvement, and Use of Inter-Firm Specific Knowledge-Sharing Routines

In this case, the analysis goes from an individual to a dyadic level and the study scope becomes the strategic alliance management block. IT tools are critical for enabling and supporting information and knowledge sharing processes among partners that allow the creation of specific routines and the possibility of obtaining relational rents (Dyer & Singh, 1998). In most cases, these routines make partners an important source of information and ideas, which result in an

Figure 2. Knowledge management systems (KMS), IT, and alliance management

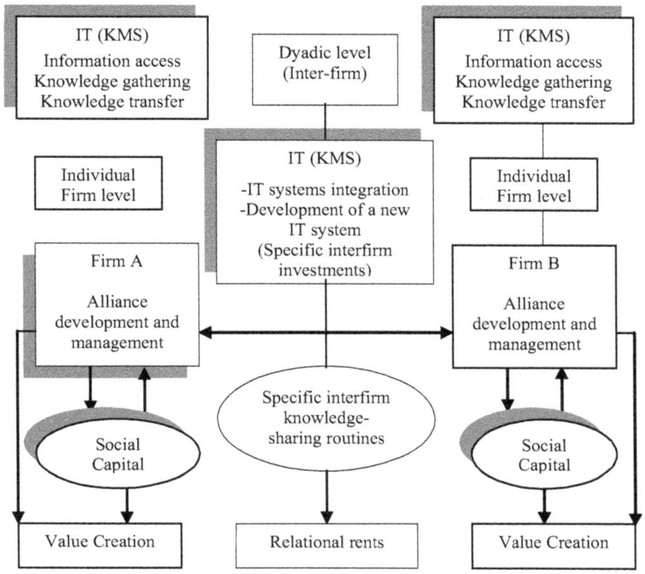

innovative stream for the firm. However, organizational differences, knowledge base, and IT structure among partners have influence on the way knowledge is managed in the alliance with the objective of leveraging the experience and competencies of the participants (Schmaltz, Hagenhoof, & Kaspar, 2004).

Figure 2 shows how IT can support the alliance development and management, within a context of a knowledge management system at firm and alliance level (considering two firms).

As the model shows, alliance advantages can arise at a two different levels: (1) individual, through value creation at the interaction dynamic process between alliances management and social capital in each firm; and (2) dyadic, through the creation and improvement of specific inter-firm knowledge-sharing routines. They can be defined as "a regular pattern of inter-firm interactions that permits the transfer, recombination or creation of specialised knowledge, (…) and these are institutionalized inter-firm processes that are purposefully designed to facilitate knowledge exchanges between alliance partners" (Dyer & Singh, 1998, p. 665).

Two issues arise in relation to knowledge transfer among alliance partners that affect IT tools and their use and effectiveness in supporting the creation and improvement of inter-firm routines and the potential for obtaining relational rents. Firstly, IT tools are required for knowledge codification processes. In doing this, knowledge can be transferred across firm boundaries and can be understood among partners—including the development of a common code shared by firms in the cooperative relationship (Kogut & Zander, 1992), which even could constitute an advantage for partners (Oliveira, 1999). In addition to this, IT tools can improve the absorptive capacity of each alliance partner, based on the support for the development of an overlapping knowledge base and the interaction between routines.

In general terms, the establishment of a coherent structure for the development of knowledge management in strategic alliances implies solving some technical problems and making decisions about what information technologies to use, along with the design and implementation of the knowledge management system.

IT Tools and their Suitability in Strategic Alliances

In relation to the appraisal of the utility and suitability of IT tools in knowledge management and in strategic alliances, it can be helpful to depart from some kind

Table 3. IT and knowledge management system (Maier, 2002)

Categories	IT tools
(1) Publication, structuring, linking	• Publication tools • Classification tools • Cross referent tools
(2) Knowledge integration	• Manual/Automatic transfer from external sources • Data evaluation, analysis, aggregation
(3) Knowledge transfer	• E-learning • Blended learning
(4) Communication and cooperation	• Message Exchange • Collaborative work
(5) Knowledge search	• Pull-search functions • Push-services • Improvements of search
(6) Knowledge presentation	• Connections between contents • Connections between search results
(7) Administration	• Reporting • User manager

of established classification due to their great diversity (Schmaltz et al., 2004). For example, Maier (2002) classifies IT tools based on the functions they serve in the knowledge management system. In this classification, tools are grouped into seven categories (Table 3).

This classification has the advantage of offering a higher degree of clarity than classifications that use only two categories based on dichotomies or strategies such as codification and personalization (see e.g., Schulz & Jobe, 2001).

The aim of this work is not to analyse the potential or suitable use of each IT tool in alliances. Rather, what is shown are some of the requirements that have to be met in order to successfully employ IT tools in cooperation agreements, based in the work of Schmaltz et al. (2004). The first group of them is determined by the particular use of IT. It may be characterized by heterogeneous systems, which may—in the worst case—be different for every partner involved, and the necessity to react to changes in the structure of cooperation, which may be frequent. Requirements are: the ability to access other systems across platforms, the flexibility of the integration, and the variability of the presentation. The other group of requirements is determined by the structure of knowledge in alliances. The main characteristic in cooperation agreements is that knowledge is distributed among participants and this leads to two requirements of IT support: the ability to bridge differences in terminology and language, and the possibility of detailed access controls.

(a) *Cross-platform accessibility.* This implies the existence of interfaces for reaching data access on the storage level between partners. Also, comprehensive knowledge management systems require interfaces for applications used to store information and knowledge. Tools used for knowledge management in alliances will have to be able to bridge technological gaps between participants (Schmaltz et al., 2004, p. 2).

(b) *Flexibility of integration.* The integration of partners' IT systems has to be as flexible as possible and must not require extensive programming efforts. This implies that even if any change in the alliance occurs (e.g., the entrance of a new partner), systems have to be incorporated quickly with no excessive efforts.

(c) *Variability of presentation.* In order to avoid problems—such as information or knowledge overload; poorly structured information or knowledge; complex or difficult information; or knowledge access—it is very important to tailor the system's content to the needs of the users.

(d) *Overcoming language differences.* Knowledge management systems should help to overcome or reduce communication problems among partners and also reduce language ambiguities. Thus, the creation of a common shared code can help to reduce these kinds of problems, as indicated previously.

(e) *Access control.* One important strategic question in alliances is the knowledge protection among partners of some modules or areas to avoid its imitation. Also, protection of the common knowledge (i.e., between partners) in relation to third parties can be important. In this case, IT in knowledge management systems must incorporate some mechanisms to control access.

The examination of the former requirements shows a choice between the use of two possible options for partners related to the investment in a common knowledge management system (as a specific asset) in the relationship. In other words, firms can integrate their own current systems—totally or partially—or to develop a specific common platform (Schmaltz et al., 2004). In this case, we would have to analyse additional aspects other than the former (i.e., cross-platform accessibility, flexibility of integration, variability of presentation, overcoming language differences, and access control) for each of the alternatives, such as cost, alliance length and type, partners' willingness, and so forth, to make the best decision.

Managing Organizational Knowledge in Strategic Alliances

Nowadays, alliances with the main objective of development and access to knowledge and technology are gaining great importance for three reasons (Peña & Aranguren, 2002): (1) they facilitate the obtaining of basic scientific inventions and discovery in basic research, reducing the typical uncertainty of R&D agreements, which allow firms to share the cost and risk of R&D projects; (2) cooperation agreements can also help to reduce the time of the product introduction into the market; and (3) alliances can be efficient tools for the marketing of the new product. A very valuable and bi-directional flow of information and knowledge can be obtained through cooperation.

Managing organizational knowledge in an alliance involves working on the best possible strategic design to create, maintain, transfer, and apply organizational knowledge developed among partners to achieve competitive goals.

There are two main aspects that can be highlighted in order to define the advantages obtained from knowledge creation, transfer, and sharing in strategic alliances: (1) fosters innovations development and (2) affords access to new knowledge.

Foster Innovations Development

In order to keep its competitive advantage, firms must foster their R&D politics. However, there are very few organizations that have available all the necessary resources and capabilities in order to attain this goal. For these reasons strategic alliances become a tool through which firms share knowledge for innovation and, in this way, they can create and sustain competitive advantages.

Firms with an efficient cooperative agreement have better access to potential information about new technological opportunities than other firms that act in an isolated manner. This leads them to achieve better results when they launch new products onto the market (Soh, 2003). On the other hand, structural flexibility is necessary for firms to adopt innovations. This can be achieved with the generation of higher levels of knowledge through cooperation agreements (Peña & Aranguren, 2002).

By way of encouraging the development and diffusion of new technologies, allowing small and medium firms to take advantage in the innovation process, a complex system of relationships between firms, universities, and public and private institutes of research must be fostered (through government actions). These relationships make easier knowledge creation; storage and transmission; and development of abilities and tools; which delimitate new technologies (Quintana & Benavides, 2003).

Access to New Knowledge

By knowledge sharing and having access to resources and capabilities that the firm lacks, development of strategies that the firm would not be able to create in isolation becomes possible. Peña and Aranguren (2002) remark that cooperation agreements are the better vehicle for the transference of tacit knowledge, abilities, and reputation.

If an agreement is developed by a company that is strongly established in the market, it is possible to gain access to valuable information and capabilities that increase the value of the firm (Gulati & Higgins, 2003). Afterward, the information will be converted into knowledge through the learning process. As has been explained previously, IT has a key role in this process. The volume and diversity of information that exists in an alliance depends on its structure, but the quality of information depends on the previous experience of alliances between partners (Koka & Prescott, 2002).

Therefore, the generation of collaborative advantages, based on knowledge exchange, depends on the interest, the absorptive capacity and collaboration of the partner that receives the knowledge, transmission and communication methods, and on the transparency of the partner that gives them. Obviously, not all the partners possess identical capabilities to learn and assimilate knowledge, and because of this there is always a risk of imbalance and opportunism. The generation of these advantages is developed in very different stages.

Phases in the Development of Alliances

Looking at knowledge generation and exploitation processes in closer detail, two stages can be differentiated in the development of the alliance (it may be that in a cooperative agreement there is only one stage, but to generate the maximum collaboration advantages it is advisable that two of them occur).

Exploration Alliances or Learning Alliances

At this stage, capability development advances more quickly and the risks of acquisition and exploitation of external knowledge are reduced (Lane & Lubatkin, 1998). The main objective is the exploration of new resources and capabilities that the firm has not obtained through cooperation agreements and that can be oriented towards: know-how acquisition and learning about different capabilities; searching of resources at partners' firms; and the creation of new specialised resources and capabilities combined with those that belong to the other firm.

Learning generated at this stage permits taking advantage of all potential benefits of R&D. Therefore, it is not possible to assure that potential learning obtained from knowledge acquisition, generation, and transfer in cooperative agreements, becomes complete. Firms have to establish processes, structures, and even develop specific politics in order to make this process easy.

Cooperation has an important role in learning and knowledge transfer processes, because the communication produced in strategic alliances boosts the creation of a common knowledge base and fosters learning between firms, improving its absorptive capacity (Lane & Lubatkin, 1998). As stated in the previous section, the application of IT is fundamental in order to obtain these advantages, because they allow knowledge storage, improve the transmission of acquired knowledge, and allow people with specific knowledge to make contact (Medina, Bruque, & Ruiz, 2003).

Trust between partners has a strong influence on learning processes and therefore, on the relationships between partners and on employee motivation. Apart from this, two kinds of learning are generated from a strategic alliance. First of all, it is important to learn the necessary knowledge for an adequate management of cooperative agreements. For this reason, previous experience is identified as an important factor in the process of partner selection because this fosters efficiency in cooperation. Also, learning about know-how, abilities, and competences of the partner firms is obtained. All of this causes some changes in the organizational routines of the firm and on the new organizational knowledge creation.

Exploitation Alliances

This stage is characterized by "the use and development of things already known" (Rothaermel & Deeds, 2004). Knowledge generated in the previous stage is jointly exploited alongside that which the firm already possesses. Sometimes, alliances only have this stage, and the objective is to take advantage of shared knowledge between firms, and not to explore new unknown fields. Positive effects for the product in the market arise and the firm has the possibility of exploiting the shared use of untapped resources and capabilities alongside those considered as basic, nowadays.

The exploration of new opportunities increases the potential benefits of all of the firm's resource exploitation due to the synergies obtained from complementary assets. The introduction of a new product can be considered as the introduction of new knowledge about a product (Madhavan & Grover, 1998), motivated by customer demand. So, the company could differentiate between nearness and quickness of response to costumer needs.

The accomplishment of the proposed goals at any of these stages is a complex process that entails many difficulties, and more difficult to resolve still, are those related to organizational knowledge management, which are analysed in the next section.

Problems for Knowledge Management in Strategic Alliances

In some cases, alliances fail or their results fall shorter than expected because there are some problems in the sharing of knowledge between firms involved. The main difficultly is to face two potential concerns: to maintain an open knowledge exchange to achieve collaborative advantages and to control knowledge flows to avoid unintended leakage of valuable technology (Oxley & Sampson, 2004).

Firms often erect barriers that make the effective development of collaborative activities, knowledge sharing, and transmission in alliances difficult. The main problems discussed in this section relate to: the specificity and complexity of knowledge, distrust between partners, the technological knowledge diversity, and the organizational form or governance structure of alliances.

The Specificity, Complexity, and Tacitness of Knowledge

In order to carry out alliance objectives successfully it is essential to share some specific knowledge between the participating partners. There are some difficulties in relation to knowledge transmission and sharing, especially when knowledge is specific, complex, and tacit.

This kind of knowledge is difficult to keep and transmit because it depends on the context, experience, language, and previous accumulated knowledge (Grant, 1996; Oliveira, 1999). Tacit knowledge is difficult to explain to others (Nonaka & Takeuchi, 1995). The same problem arises with knowledge that is specific to a context or culture (Zander & Kogut, 1995). Transferring tacit and specific knowledge is a very costly and lengthy process. Joint ventures are used as a governance mechanism when the alliance implies sharing especially tacit and complex knowledge, because results and the process are very difficult to specify and monitor in these kinds of agreements (Oxley & Sampson, 2004, p. 725).

When knowledge is explicit, it is easier to share and to transfer, but it has a shorter strategic value than tacit knowledge (Zander & Kogut, 1995). Acquiring the latter kind of knowledge can be the main reason to participate in an alliance.

Protection of Strategic Knowledge

One of the main risks of strategic alliances is the difficultly in protecting some essential knowledge (Oxley & Sampson, 2004, p. 727): hints and ideas about strategy orientation, directions, and partial results of technological research; competitive benchmarking data; codified knowledge contained in formulas, design, and procedures; tacit knowledge involved in skills and routines; and the essential competences of key employees that can be hired when the alliance is finished.

The risk of the partner developing an opportunistic behaviour in the alliance and appropriating this kind of strategic knowledge is one of the main factors that discourage firms from participating in an alliance.

Mutual Distrust between Partners

Mutual distrust between partners of an alliance is another factor that makes learning in alliances and knowledge sharing difficult. Firms are especially reluctant to share knowledge with companies that may become competitors in the future.

Mechanisms and systems to reduce distrust and opportunism and to improve collaboration between partners are very important. The common space created in an alliance allows partners to share their tacit and explicit knowledge and their abilities and productive processes. Furthermore, if an adequate level of trust is attained it is possible to create and exchange new knowledge, especially that which is tacit, generated in the agreement and which afterwards is absorbed and assimilated by firms, leading them to improve their capabilities (Inkpen, 1998). Therefore, it is possible to affirm that the last goal of cooperation, carried out to access the partner's knowledge, is the internalisation of this knowledge and also of that generated in the alliance. In this case, cooperation is a tool that makes organizational learning easier.

Technological and Knowledge Diversity

Another important factor in arriving at a collaborative advantage is the technological and knowledge similarity of the partners (Kwak, 2004), defined as the degree to which the partners' portfolios overlap (Oxley & Sampson, 2004). If the technological level of partners is too different, they have very little knowledge and few capabilities to share. However, when it is too similar, the transmission of it is of no interest or value. In this case, they can't learn anything from one another, because they have no strategic knowledge to share. There can be moderate differences in technological capabilities between partners in an alliance (Oxley & Sampson, 2004).

Firms are chosen for a cooperative agreement in relation to the potential value of their knowledge. The ideal situation is when knowledge is complementary and compatible (Inkpen, 1998), that is to say, when the firm has basic compatible knowledge, but at the same time, knowledge that is of a different specialised nature. Knowledge heterogeneity fosters creativity and innovation in the firm and also improves firm performance.

A very closely related aspect to common knowledge that a partner can develop is their absorptive capacity (Cohen & Levinthal, 1990), which is a function of (1) the extent to which partners have developed an overlapping knowledge base,[4] and (2) the extent to which partners have developed interaction routines that maximize the frequency and intensity of socio-technical interactions (Dyer & Singh, 1998, p. 665). Owing to this, the main problem for knowledge transferring and sharing between partners can arise when there is not a common knowledge base between them. A certain redundancy is desirable because knowledge, or information meaning, may be ambiguous (Lane & Lubatkin, 1998; Nahapiet & Ghoshal, 1998).

Alliance Organizational Form or Governance Structure

An alliance's organizational form is another very important factor related to the achievements that the partners can attain (Oxley & Sampson, 2004). The lack of common routines, clear lines of authority, and little hierarchical organizational forms can make cooperation difficult, along with sharing of knowledge and learning of the members of the alliance. Although all organizational forms have disadvantages, partners must implement that which is optimal according to the nature of the agreement (joint ventures, alliances, equity joint ventures, etc.). The more hierarchical the organizational forms are, the more they facilitate the control of processes and performance; however, they imply higher costs and more bureaucracy (Oxley & Sampson, 2004).

The governance structure or the organizational form is also an important mechanism in achieving a balance between the exchange and protection of knowledge flows in alliances (Kale et al., 2000; Oxley, 1997). This is especially difficult when firms are direct competitors in products and/or strategic resources.

Cultural Barriers to Collaboration

Cultural barriers to collaboration are often difficult to detect and solve but have a strong influence on alliance success. One of the most important is the culture of firms that promote and reward individualism, as employees are then reluctant to share what they know because they have no incentive to do so (Hansen & Nohria, 2004, p. 25).

Another cultural barrier arises when leaders and employees consider that time devoted to communication and collaboration with other members is not productive. This lack of communication makes it difficult to identify both problems and the experts who have specific knowledge and are able to resolve those problems.

Solutions for Organizational Knowledge Problems

Some solutions are proposed to resolve the problems of sharing and transferring knowledge in alliances. To achieve solutions for organizational knowledge management and arrive at the development of collaboration advantages some key factors must be considered: the generation of an optimal environment for collaboration; training; previous experience; organizational culture and willingness to cooperate; the level of trust; the capability of transforming tacit into explicit knowledge; and the capability of learning from the firms that cooperate.

First of all, members must understand the potential performance of the alliance and be willing to share knowledge-based capabilities. To avoid the risk of the partner developing opportunistic behaviour—inherent in all cooperative agreements—the establishment of formal agreements on the resource contributions necessary for an adequate working of the agreement are very important. Also, the risks of strategic alliances reduce when partners have previously had positive experiences in cooperation. In this case, a higher value will be generated in future alliances (Stuart, 2000).

In relation to technological and knowledge diversity it is important to elect a partner with which the firm shares some similarities in terms of size, previous experience in cooperation, organizational culture, financial resources, or work climate (Rothaermel & Deeds, 2004; Saxton, 1997). The greater the similarities between the partners' firms, the more agile relationships will be between them and the stronger the trust that will be generated in the agreement (Saxton, 1997). This contributes to achieving higher efficiency in the accomplishment of the agreement goals. Mowery, Oxley, and Silverman (1998) highlight the importance of knowledge complementarities and partner-specific absorptive capacity for the success of strategic alliances.

To resolve problems related to transfer, tacit knowledge must be modified (i.e., codified as much as possible) and incorporated into particular conditions of the partner. IT tools can be used as an important aid to try to transmit knowledge. However, some organizational measures and human resources practices, such as cross-firm work teams, fostering open communication or the reward systems, are essential to motivate the sharing and transferring of tacit knowledge (Guadamillas, 2001b; Oxley & Sampson, 2004).

In order to resolve organizational problems, alliances must have formal governance, and as much as possible, organizational independence. In addition to this, mechanisms and systems that facilitate knowledge sharing and transmission, such as work teams, can be useful to eliminate barriers in cooperation. Pisano, Russo, and Teece (1998) highlight the adoption of hierarchical and protective governance structures, especially for equity joint ventures in strategic alliances.

For wide alliance scope, defined as the number of technologies of functional activities involved (Pisano, 1989), hierarchical governance structures are more suitable, because they provide clear lines of authority and common routines. These can be important factors in reinforcing partner trust to ensure that knowledge cannot be captured in an opportunistic form.

Human resource practices such as incentives; promotion; selection and recruitment systems; and work teams can help to resolve cultural problems related to the rejection of collaboration. Development of strong relationships between employees of firms involved in the alliance and cross-firm work teams make knowledge transmission easier. Also, benchmark systems help people to identify and learn best practices from the partner that are needed to work together in accomplishing alliance objectives.

IT, as a database, can facilitate communication and knowledge transmission. Other solutions proposed by Hansen and Nohria (2004, p. 27) are the strong support of leaders to collaborate in alliances by working with other partners; developing unifying goals and motivating people to achieve them; the creation of information networks; and development of unifying goals to facilitate knowledge transmission in alliances.

One of the main concerns of firms in strategic alliances is the protection of strategic knowledge that has to be increasingly shared according to the development of the agreement. To solve this problem, Reuer, Zollo, and Singh (2002) propose a reduction in the scope of activities involved in R&D agreements.

All the solutions proposed are especially difficult to implement when firms are direct competitors in terms of products and resources. In this case, Oxley and Sampson (2004) propose two solutions with a narrow relationship in order to protect knowledge resources in R&D alliances: limit the scope of alliance activities and carefully regulate knowledge sharing. Decisions about alliance scope can be some of the most important in strategic alliances. They involve defining boundaries for groups of products, brands, fixed assets, or activities in which partners participate. Finally, the regulation of knowledge sharing involves the use of organizational forms and governance structures in strategic alliances.

Discussion and Conclusions

The main objective of this chapter has been to explain the generation of collaborative advantages in strategic alliances. Through cooperation agreements firms can create and maintain a competitive advantage because they provide, for example, the possibility for firms to access complementary resources and capabilities; to enter into new markets; or to improve learning. However, the adequate sharing and transmission of knowledge between partners is an essential requirement to achieve this advantage. Problems involved in these processes and the role of information technologies have been addressed in previous sections.

Theoretical perspectives of resources, knowledge, and social capital recognize the importance of strategic and social questions in establishing strategic alliances. The more common economic approaches used to analyse this subject (i.e., transaction cost theory) are not able to adequately explain these issues. Moreover, some important questions related to the problems and advantages of the knowledge sharing process and the use of IT tools in strategic alliances have been explained by these approaches.

It is important to comment on the main conclusions reached in reference to the role of IT tools in strategic alliances management. On the one hand, there are not any specific IT tools for alliance management, but in any case, those which are more suitable must be chosen according to the specifics of the alliance, especially those related to knowledge transmission between partners. Not only technical but also organizational and human aspects must be considered in making this choice. On the other hand, IT tools are very useful for knowledge transfer, based on the ability of the firm to codify and establish a common code or language. Moreover, the existence of an interaction between social capital, alliance management, and value creation make the development process of agreements dynamic and very difficult to separate into blocks of differentiated problems—previous aspects and management activities of the alliance.

All the specific organizational aspects of cooperation agreements, such as limitations for the use of hierarchical coordination, the need of motivation between partners for participation, and the investment in knowledge management activities, alongside all the technical problems, such as the technological resource availability; the system of knowledge management design and implementation; and the content problems, such as more adequate technology which needs to be applied, are all complex questions without an easy solution. These make the establishment of a coherent structure for knowledge management in strategic alliances a very complicated question indeed (Schmaltz et al., 2004).

Some important problems that arise in strategic alliances make the effective development of collaborative activities and knowledge sharing and transmission in alliances difficult. The first problem is the specificity and tacitness of knowledge that leads to its effective storage and transmission being more difficult. Moreover, the distrust between partners and the cultural barriers in relation to collaboration cause partners to be reluctant to participate in alliances. The excessive technological and knowledge diversity or similarity can be a problem for learning from partners. Finally, the alliance organizational form or governance structure must be adapted to alliance objectives and the firms' specific requirements.

To resolve some of the problems in strategic alliances, several solutions have been proposed: the use of IT tools, organizational measures, human resource practices, and the limitation of the scope of the agreement.

Finally, it is important to consider all the factors and solutions explained, as nowadays, advantages based on traditional scale and scope economies are rapidly diminishing. Alliances will lead firms to build a competitive advantage with others that they cannot build alone in the face of their rivals. The most important element will be the efficient transmission and sharing of knowledge that will allow companies to gain and maintain competitive advantages over their competitors.

References

Adler, P., & Kwon, S. W. (2002). Social capital: Prospects for a new concept. *Academy of Management Review, 27,* 17-40.

Arjonilla, S. J., & Medina, J. (2002). *La gestión de los sistemas de información en la empresa.* Madrid: Pirámide.

Barney, J. B. (1991). Firm resources and sustained competitive advantage. *Journal of Management, 17*, 99-120.

Bowman, C., & Faulkner, D. (1997). *Competitive and corporate strategy.* London: Irwin.

Chung, S., Singh, H., & Lee, K. (2000). Complementary assets, status similarity and social capital as drivers of alliance formation. *Strategic Management Journal, 21,* 1-22.

Clemons, E. K., & Row, M. C. (1992). Information technology and industrial cooperation: The changing economies of coordination and ownership. *Journal of Management Information Systems, 9*(2), 9-28.

Cohen, W. M., & Levinthal, D. A. (1990). Absorptive capacity: A new perspective on learning and innovation. *Administrative Science Quarterly, 35,* 128-152.

Dyer, J. H., & Singh, H. (1998). The relational view: Cooperative strategy and sources of competitive advantage. *Academy of Management Review, 23,* 660-679.

Eisenhardt, K. M., & Schoonhoven, C. B. (1996). Resource-based view of strategic alliance formation: Strategic and social effects in entrepenurial firms. *Organization Science, 7*(2), 136-150.

Foss, N. (1999). Research in the strategic theory of the firm: Isolationism and integrationism. *Journal of Management Studies, 36*(6), 725-755.

Grant, R. M. (1996). Toward a knowledge based theory of the firm. *Strategic Management Journal, 17,* 109-122.

Gulati, R. (1998). Alliances and networks. *Strategic Management Journal, 19,* 293-317.

Gulati, R., & Higgins, M. C. (2003). Which ties matter when? The contingent effects of interorganizational partnerships on ipo success. *Strategic Management Journal, 24,* 127-144.

Gulati, R., Nohria, N., & Zaheer, A. (2000). Strategic networks [Special issue]. *Strategic Management Journal, 21,* 203-215.

Guadamillas, F. (2001a). La gestión del conocimiento como recurso estratégico en un proceso de mejora continua. *Alta Dirección, 217,* 199-209.

Guadamillas, F. (2001b). La gestión del conocimiento desde una perspectiva basada en los recursos. Análisis de las implicaciones para la dirección de recursos humanos [nº especial de recursos humanos]. *Revista del centro de estudios financieros de trabajo y seguridad social, 216,* 115-154.

Hagedoorn, J. (1993). Understanding the rationale of strategic technology partnering: Interorganizational modes of cooperation and sectoral differences. *Strategic Management Journal, 14,* 371-385.

Hagedoorn, J., & Schakenraad, J. (1994). The effect of strategic technology alliances on company performance. *Strategic Management Journal, 15,* 291-309.

Hamel, G. (1991). Competition for competence and interpartner learning within international strategic alliances. *Strategic Management Journal, 12,* 83-103.

Hansen, M. T., & Nohria, N. (2004, Fall). How to build collaborative advantage. *Sloan Management Review, 46*(1), 22-30.

Hitt, M. A., Ireland, R. D., & Santoro, M. D. (2004). Developing and managing strategic alliances, building social capital and creating value. In A. Ghobadian, N. O'Reagan, D. Gallear, & H. Viney (Eds.), *Strategy and performance: Achieving competitive advantage in the global marketplace* (pp. 13-14). New York: Palgrave.

Inkpen, A. C. (1998). Learning and knowledge acquisition through international strategic alliances. *Academy of Management Executive, 12*(4), 69-80.

Inkpen, A. C., & Beamish, P. W. (1997). Knowledge, bargaining power, and the instability of international joint ventures. *Academy of Management Review, 22*(1), 177-202.

Ireland, R. D., Hitt, M. A., & Vaidyanath, D. (2002). Alliance management as a source of competitive advantage. *Journal of Management, 28,* 413-446.

Kale, P., Singh, H., & Perlmutter, H. (2000). Learning and protection of proprietary rights in strategic alliances: Building relational capital. *Strategic Management Journal, 21,* 217-237.

Kogut, B. (1988). Joint-ventures theoretical and empirical perspectives. *Strategic Management Journal, 9,* 319-322.

Kogut, B., & Zander, U. (1992). Knowledge of the firm, combinative capabilities and the replication of technology. *Organization Science, 3,* 383-397.

Koka, B. R., & Prescott, J. E. (2002). Strategic alliances as social capital: A multidimensional view. *Strategic Management Journal, 23,* 795-816.

Koza, M. P., & Lewin, A. Y. (1998). The co-evolution of strategic alliances. *Organization Science, 9*(3), 255-264.

Kwak, M. (2004, Fall). Maximizing innovation in alliances. *Sloan Management Review, 46*(1), 5-6.

Lane, P. J., & Lubatkin, M. (1998). Relative absorptive capacity and interorganizational learning. *Strategic Management Journal, 19*(5), 461-477.

Lee, C., Lee, K., & Pennings, J. M. (2001). Internal capabilities, external networks and performance: A study on technology-based ventures. *Strategic Management Journal, 22,* 615-640.

Liebeskind, J. P. (1996). Knowledge, strategy and the theory of the firm. *Strategic Management Journal, 17,* 93-107.

Madhavan, R., & Grover, R. (1998). From embedded knowledge to embodied knowledge: New product development as knowledge management. *Journal of Marketing, 62,* 1-12.

Maier, R. (2002). *Knowledge management systems: Information and communication technologies for knowledge management.* Berlin: Springer.

Medina, J. A., Bruque, S., & Ruiz, J. (2003). *Evidencias empíricas de cómo las tecnologías de la información estimulan las redes estratégicas.* Salamanca: XIII Congreso Nacional de ACEDE.

Menguzzato, M. (1995). La triple lógica de las alianzas estratégicas. In A. Cuervo (Ed.), *Dirección de empresas de los noventa. Homenaje al profesor Marcial Jesús López Moreno.* Madrid: Civitas.

Montoro, M. A. (2000). *Factores determinantes del éxito de la cooperación entre empresas. Una aplicación en acuerdos internacionales en investigación y desarrollo.* Oviedo: X Congreso Nacional de ACEDE.

Mowery, D. C., Oxley, J. E., & Silverman, B. S. (1998). Technological overlap and interfirm cooperation: Implications for the resource—based view of the firm. *Research Policy, 27,* 507-523.

Nahapiet, J., & Ghoshal, S. (1998). Social capital, intellectual capital and the organizational advantage. *Academy of Management Review, 23,* 242-266.

Navas, J. E., & Guerras, L. A. (2002). *La dirección estratégica de la empresa: Teoría y aplicaciones.* Madrid: Civitas.

Nohria, N., & Garcia-Pont, C. (1991, Summer). Global strategic linkages and industry structure [Special issue]. *Strategic Management Journal, 12,* 105-124.

Nohria, N., & Piskorski, M. J. (1997). *Focus and diversification: The effects of changes in the scopes of large corporations.* Working paper. Boston: Harvard Business School.

Nonaka, I., & Takeuchi, H. (1995). *The knowledge creating company.* New York: Oxford University Press.

Oliveira, M. (1999). Core competencies and the knowledge of the firm. In M. A. Hitt, P. Gorman, R. D. Nixon, & K. P. Coyne (Eds.), *Dynamic strategic resources: Development, diffusion and integration* (pp. 17-41). New York: John Wiley & Sons.

Oxley, J. E. (1997). Appropriability hazards and governace in strategic alliances: A transaction cost approach. *Journal of Law, Economics and Organization, 13*(2), 387-409.

Oxley, J. E., & Sampson, R. C. (2004). The scope and governance of international R&D alliances. *Strategic Management Journal, 25,* 723-749.

Parkhe, A. (1998). Messy research, methodological predispositions and theory development in international joint ventures. *Academy of Management Review, 18*(2), 227-268.

Peña, I., & Aranguren, M. J. (2002). Transferencia de conocimiento mediante acuerdos de colaboración. *Economía Industrial, 346,* 67-80.

Peteraf, M. (1993). The cornerstones of competitive advantage: A resource-based view. *Strategic Management Journal, 14,* 179-191.

Pfeffer, J., & Salancik, G. (1978). *The external control of organizations: A resource dependence perspective.* New York: Harper & Row.

Pisano, G. (1989). Using equity participation to support exchange: Evidence from the biotechnology industry. *Journal of Law, Economics and Organizations, 5*(1), 109-126.

Pisano, G., Russo, M., & Teece, D. (1988). Joint ventures and collaborative arrangements in the telecomunications equipment industry. In D. Mowery (Ed.), *International collaborative ventures in U.S. manufacturing* (pp. 23-70). Cambridge, MA: Ballinger.

Quintana, C., & Benavides, C. A. (2003). *Concentraciones geográficas, innovación y conocimiento tecnológico.* Salamanca: XIII Congreso Nacional de ACEDE.

Reuer, J. R., Zollo, M., & Singh, H. (2002). Post-formation dynamics in strategic alliances. *Strategic Management Journal, 23*(2), 135-151.

Rialp, J. (1999). Experiencia de las empresas españolas con los acuerdos de cooperación. *Revista Icade, 48,* 149-165.

Rothaermel, F. T., & Deeds, D. L. (2004). Exploration and exploitation alliances in biotechnology: A system of new product development. *Strategic Management Journal, 25,* 201-221.

Sanchez, R., & Mahoney, J. T. (1996). Modularity, flexibility and knowledge management in product and organization design. *Strategic Management Journal, 17,* 63-76.

Saxton, T. (1997). The effects of partner and relationship characteristics on alliances outcomes. *Academy of Management Journal, 40*(2), 443-461.

Schmaltz, R., Hagenhoff, S., & Kaspar, C. (2004, April). *Information technology support for knowledge management in cooperation.* Paper presented at the Fifth European Conference on Organizational Knowledge, Learning and Capabilities, Innsbruck, Austria.

Schulz, M., & Jobe, L. A. (2001). Codification and tacitness as knowledge management strategies: An empirical exploration. *Journal of High Technology Management Research, 12,* 139-165.

Soh, P. H. (2003). The role of networking alliances in information acquisition and its implications for new product performance. *Journal of Business Venturing, 18,* 727-744.

Spender, J. C. (1996). Making knowledge the basis of a dynamic theory of the firm. *Strategic Management Journal, 17,* 45-62.

Stuart, T. E. (2000). Interorganizational alliances and the performance of firms: A study of growth and innovation rates in a high-technology industry. *Strategic Management Journal, 21,* 791-811.

Williamson, O. (1975). *Markets and hierarchies: Analysis and antitrust implications.* New York: Free Press.

Zander, U., & Kogut, B. (1995). Knowledge and the speed of transfer and imitation of organizational capabilities: An empirical test. *Organization Science, 6,* 76-92.

Endnotes

1 Knowledge management can be understood as a dynamic cycle of processes leading to the generation, storage, transfer, and application of knowledge with the objective of innovation and continuous improvement in the organization (Guadamillas, 2001a).

2 This work highlights the importance of social capital at an individual level. From this perspective, it can be understood that it helps a firm to reach out the access to resources through social relationships networks (Ireland et al., 2002). Also, social capital facilitates the development of intellectual capital within the firm by its influence in the conditions that allow the combination and exchange of knowledge to happen (Nahapiet & Ghoshal, 1998).

3 For an overview about social capital from several disciplines and an assessment of its utility for organizational theory, see Adler and Kwon (2002).

4 It does not mean that their knowledge bases must be similar. Instead, some common basic knowledge among firms that allow inter-organizational knowledge to occur must exist (Lane & Lubatkin, 1998).

Chapter VI

The Repercussions of the Adoption of Information Technology on Co-Operative Innovation

Cristina Bayona-Sáez, Universidad Pública de Navarra, Spain

Teresa García Marco, Universidad Pública de Navarra, Spain

Abstract

The aim of this chapter is to analyse what repercussion the adoption of certain types of information and communication technology (ICT), by firms, may have on collaborative innovation. It is our belief that tools such as Internet access, e-mail, and the presence of internal and inter-organisational communication networks, will promote and facilitate the signing of cooperative agreements. The empirical analysis performed on a sample of 1,649 firms partially confirms this intuition, since the existence of a network, bringing the firm into contact with its subcontractors, customers, and/or suppliers, increases the propensity of firms to cooperate in innovation.

Introduction

Far from being a recent phenomenon, the expansion of ICT goes back at least 50 years (Salas, 2001). It is only in the last 15 years, however, that ICT has acquired its current status as the focus of numerous studies. Due to the evolution of network technologies and the decrease in transaction costs[1] through ICT, (Malone, Yates, & Benjamin, 1987) new opportunities have arisen for businesses to broaden their scope and carry out exchanges worldwide (Kelly, 1998). As a result, ICT is seen as one of the main drivers of economic growth, although in this respect, Europe in general, and Spain in particular, follow the lead of the United States. In 1999, per capita ICT expenditure in Europe was 60% that of the United States (European Commission, 2001).

Collaboration in innovation between firms is also anything but a new phenomenon. The literature on innovation indicates that the last two decades have seen systematic change in the way firms undertake innovation activities (Narula, 2001). The use of external networks, in particular, has grown significantly (Hagedoorn, 1996). Indeed, Duysters, Kok, and Vaandrager (1999) report that alliances are no longer regarded as peripheral features, but as one of the mainstays of the firm's technological strategy. In addition to the declining costs of monitoring and exploiting networks, a growing need has arisen for firms to possess multiple technological competencies (Granstrand, Patel, & Pavitt, 1997). Firms have sought to utilise *non-internal* means, such as strategic alliances and outsourcing, to undertake innovation (Hagedoorn 1996; Narula & Hagedoorn 1999; Tidd & Trewhella 1997). In this respect also, Europe and Spain lag behind the United States and Japan.

This chapter aims to describe and analyse the repercussions of the adoption of ICT on collaborative innovation. Awareness of the issues surrounding ICT and collaborative innovation has led the various national administrations, and specifically the European Union, to implement a variety of instruments to foment them both. One of the priority themes of the Sixth Framework Programme is that of the information society technologies, which depend essentially on the promotion of innovation and technology transfer in Europe and the creation of technological enterprises (European Commission, 2000).

All this reinforces our belief in the relevance of this topic and in the need to explore the impact of the adoption of the new technologies on cooperation and innovation. This chapter aims to shed some light on the subject by means of an empirical analysis conducted on a sample of 1,649 Spanish manufacturing firms. These are innovating firms who replied to the questionnaires sent out by the Instituto Nacional de Estadística (National Institute of Statistics) for their surveys *Encuesta sobre Innovación Tecnológica en las Empresas (Survey of Innovation Technology in Firms) 1998*[2] and *1996*. Given that the adoption of ICT can vary across sectors, depending on firm size and the type of knowledge to be transmitted, the analysis is repeated, with the sample segmented by size and technological intensity.

The chapter continues with a section defining two of the basic concepts required for a proper understanding of what follows; namely, innovation and collaboration. It also offers a description of the main firm characteristics reported in the existing literature as having an effect on cooperation in innovation. The third section presents the theoretical framework used to explain the potential impact of the different kinds of ICT on cooperative innovation and the degree to which they have been adopted in European countries. The fourth contains an empirical analysis conducted on a sample of 1,649 Spanish manufacturing firms. Finally, the main conclusions are discussed in the final section.

A Definition of Some Key Concepts: Innovation and Collaborative Innovation

Innovation in the Firm

We will begin this section with a number of definitions that appear in the Organisation for Economic Cooperation and Development (OECD) Oslo Manual (2002).

Technological innovations include not only new products and processes, but also significant technological improvements in existing ones. A product is considered to be an innovation once it has been introduced on the market, and a process is considered an innovation when it has been used in a production process. As well as technological innovation, there also exists the term organisational innovation, in recognition of the fact that competitiveness may be increased by the implementation of new methods in the organisation of work in areas such as labour management, distribution, finance, or manufacturing. This term also covers innovation in business models.

There are various types of product innovations. On the one hand, there are complete product innovations, which differ significantly from existing products with respect to their intended use, specifications, technological characteristics, properties, and so forth. They may involve new technology or a combination of new and existing technologies. An upgraded product, on the other hand, is an existing product whose performance has been significantly enhanced. As well as product innovations, we can also talk of process innovations—the adoption of technologically new or significantly improved production methods—that may involve changes in equipment and/or in the organisation of production.

An innovating firm, therefore, according to the Oslo Manual, is one that has successfully introduced a technological innovation. This is a rather restrictive definition, since it requires firms to have been successful in their innovative process. However, the novel aspect of the innovation may not only be for the firm or group of firms; it may also be for the operating market in general. Thus, the definition of innovation comes to include not only the firm that first introduces the innovation on the market, but also those that follow its lead.

Innovation is a complex, diversified activity with many interacting components, the outcome of which is often uncertain. It includes such activities as the generation of knowledge new to the firm (through in-house R&D, the construction and testing of a prototype, and the development of certain kinds of software); the acquisition of technological know-how (through the acquisition of patents, licences, trade marks, machinery, and technology); industrial engineering; industrial design; training of personnel in new techniques; and marketing for technologically new or improved products.

The firm's innovation activities may take place in-house, be acquired from outside sources, or through collaboration with other firms. The greater the firm's chances of appropriating innovation benefits, the greater its motivation to undertake in-house innovation. The appropriation of innovation benefits has to do with the kind of knowledge involved[3] (tacit or explicit) and the effectiveness of legal provisions (such as patents) to protect inventions. It should be remembered also that the firm may carry out all the aforementioned innovation activities on its own. This requires large amounts of assets or resources, which may not be available to all firms.

Great effort is currently underway in the European Union to promote the innovation performance of firms to the maximum. In fact, improvement in innovation is one of the key aspects of the strategy to meet the target agreed at the March 2000 meeting of the European Council in Lisbon, which was to make the Union the most competitive and dynamic knowledge-based economy by the end of the decade. According to the Innobarometer (2004), research continues to be the top priority of one in two European enterprises recently involved in innovative activities. However, it is more common for these firms to conduct in-house research (53%) than to outsource it to other firms, universities, or research institutes (20%). Table 1 shows the country breakdown of results for companies conducting in-house research and those outsourcing research to other firms, universities, or research institutes. In-house research seems to be particularly prevalent in Italy (75%), Finland (74%), and the Netherlands (71%). In all *member states*, with the exception of Finland (51%), only a minority of enterprises outsource research.

Table 1. In-house vs. outsourcing research (Innobarometer, 2004)

	Q1. In the last two years, did your firm... ?		
	Sample	a) Carry out in-house research	b) Outsource research to other firms, universities, or research institutes
EU (25 countries)	4534	53 %	20 %
EU (15 countries)	3021	54 %	20 %
Belgium	208	66 %	31 %
Denmark	209	24 %	7 %
Germany	300	48 %	20 %
Greece	100	52 %	14 %
Spain	300	46 %	22 %
France	300	53 %	17 %
Ireland	99	64 %	21 %
Italy	300	75 %	16 %
Luxembourg	100	59 %	10 %
Netherlands	200	71 %	42 %
Austria	200	58 %	19 %
Portugal	100	24 %	11 %
Finland	101	74 %	51 %
Sweden	204	25 %	27 %
United Kingdom	300	63 %	19 %

Other data supplied by the Innobarometer 2004 indicate that in-house research is more common in large enterprises and/or industrial firms. The reason for this, as might be expected, is that large firms are better endowed in terms of resources and funding.

As we are about to explain, when the firm is lacking in motivation or the necessary resources to undertake in-house innovation, the solution may be to collaborate with other firms.

Collaborative Innovation

Alliances are an increasingly frequent form of corporate governance, which according to Teece (1986, 1992, 1996), are suitable for innovation in many cases because they solve some of the problems that typically arise, such as uncertainty, appropriation of research results, dissemination of tacit knowledge, or the extra assets required to follow the innovation process right through to the launching of the product on the market.

Any kind of business activity can involve a collaborative agreement; however, we should perhaps define what we mean by collaboration. One way to define it is as an agreement between two or more independent organisations to pool or share some of their skills and resources, without reaching an actual merger, and interrelate, to some extent, in order to improve their competitive advantage.

Although collaboration can take place in any of the firm's activities, we will focus in this chapter on agreements for collaborative innovation. Given the circular or interactive nature of the innovating process, firms are experiencing an ever-growing need to collaborate with other organisations in order to carry out innovation activities. Innovation sources are not to be found exclusively within firms; they are also present in other organisations and the more interaction a firm has with other organisations, the easier it will learn of new opportunities. Among the organisations who may offer complementary resources are competitors, suppliers, customers, research centres, or universities.

Collaborative innovation is defined by the Spanish National Institute of Statistics, our data source, as the active participation in joint R&D and innovation projects with other institutions and those of the firm's own projects that are officially linked to other institutions. Merely contracting outside the firm, when the firm plays no active part, is not considered collaboration. Innovation is costly and uncertain; it calls for specific managerial skills and the results it produces are difficult to appropriate (Kay, 1993). This explains the need for inter-firm collaboration in innovation.

Finally there are a set of firm characteristics, relating to size, R&D capacity, and the complexity of technological development that can influence propensity to co-operate in innovation.

R&D Capacity

One issue that the literature approaches from a variety of perspectives is the relationship between R&D expenditure in firms and R&D collaboration between firms. The absorptive capacity theory, however, tells us that a firm wishing to access and make use of the knowledge of a partner first needs to possess the necessary absorptive capacity. A firm's absorptive capacity is based on its ability to recognise the value of new information, assimilate it and apply it for commercial purposes. According to Cohen and Levinthal (1990), a firm must have prior knowledge of a subject in order to be able to assimilate and absorb further knowledge. The firm's absorptive capacity therefore depends on its R&D investment, which fulfils a twofold purpose. It is not only a means to generate new knowledge, but also to develop the firm's absorptive capacity. This determines the extent to which it is able to assimilate knowledge generated by outside sources, through collaborative agreements. The absorptive capacity theory has been tested in a number of different studies with mixed results. Cohen and Levinthal (1990) confirm the importance of absorptive capacity using observations on business units, while Griffith, Redding, and Van Reenen (2000) use a panel of industries across 13 OECD countries, and Girma (2002) used individual firms. Various studies conclude that independent R&D increases the firm's propensity to engage in cooperative R&D (Cohen & Levinthal, 1989, 1990; Colombo & Garrone, 1998). A large number of the empirical studies on this subject confirm this idea (Arora & Gambardella, 1994; Bayona, García-Marco, & Huerta, 2001; Colombo & Garrone, 1998; Fritsch & Lukas, 2001; Rocha, 1997). In contrast, Kleinknecht and Van Reijnen (1992) find no evidence to confirm Cohen and Levinthal's (1989) hypothesis.

Firm Size

According to Robertson and Gatignon (1998), there is a contradiction in the theoretical arguments relating to the impact of firm size on the decision to collaborate in R&D. While Patel and Pavitt (1994) report that the technological diversity of larger firms gives them easier access to complementary technologies, Rocha (1997) finds them more likely to invest in the sunk costs associated with technology because they are better able to obtain the necessary funding. A firm's size will therefore influence the decision to enter into collaborative innovation to a greater or lesser extent, depending on its absorptive capacity: small firms, with less capacity to undertake in-house R&D, will be less able to access and apply knowledge from a partner. In any event, most of the empirical evidence suggests that there is a positive relationship between size and cooperation in innovation (Bayona et al., 2001; Cassiman & Veugelers, 2002; Colombo & Garrone, 1998; Fritsch & Lukas, 2001; Hagedoorn & Schakenraad, 1994).

The Complexity of Technological Development

Arguments have been put forward from perspectives as varied as the transaction costs theory, the organisational theory, or the absorptive capacity theory, suggesting that the frequency of the decision by firms to collaborate in innovation increases with the technological complexity[4] of the operating sector.

The transaction costs theory, an extension of Coase's (1937) argument relating to the interpretation of *market* and *firm* as alternative mechanisms of resource allocation, received its definitive boost in Williamson (1975). According to this approach, the use of the pricing mechanism in the allocation of resources (market) involves costs, deriving from the need to negotiate, formalise and enforce the contracts corresponding to the transactions. The allocation of resources within firms is not done through the market mechanism but through a hierarchy, which also involves costs, such as those deriving from employment contracts and the difficulty of establishing an appropriate incentive system. The birth of the firm is precisely that point at which it becomes less costly to carry out certain transactions through the hierarchy than through the market. Between these two extremes there emerge other intermediate governance structures aimed at reducing transaction costs, such as agency contracts, franchises, industrial associations, and cooperative R&D, among others.

Robertson and Gatignon (1998) report that a transaction involving complex technology acquires contractual complexity. Thus, as existing technology grows in complexity, the market becomes a less efficient environment for firms to acquire the knowledge and technology to suit their needs. The contractual complexity of such transactions on the market is extreme, due to asymmetry of information between buyer and seller. Technological activities, however, are both costly and difficult to perform in-house because they call for specific capabilities. As technological complexity increases, therefore, more firms are opting to collaborate (Shing, 1997).

Organisational theories, meanwhile, stress the importance of the environment in firm performance. Since no organisation is able to generate all the resources it needs, it must obtain them from the environment. There is a wide variety of resources: raw materials, capital, personnel, technological innovations, service, and so forth. These resources are acquired from other organisations with whom exchanges are necessary in order to ensure survival. There will be less uncertainty involved in the exchange relationships that evolve between firms if they are united in a cooperative relationship.

One of the motives suggested in the organisational literature for taking part in an alliance is that it enables firms to acquire and internalise the capabilities and skills of their partners, which they can use to create new skills to suit their own needs (Cassiman, 1999; Hamel 1991; Santangelo, 2000; Steensma, 1996). In environments of uncertainty and high technological complexity, firms are able, through cooperation, to retain their capabilities and knowledge in a given activity, while acquiring new ones that will open up options and opportunities in new markets and activities in the future (Kogut, 1988; Kogut & Zander, 1992). Firms operating in hi-tech sectors will show greater absorptive capacity as a result of their own stock of technological experience and scientific capabilities. These firms will be in the best position to make effective use of cooperative knowledge output.

All these arguments suggest that inter-firm cooperation in R&D will be more frequent in technologically more complex sectors.

ICT on Cooperation in Innovation

To the aforementioned firm characteristics, already widely documented in the literature, we add one more; namely, the utilisation of new ICT. Our aim in this chapter is to test the influence of these new technologies on the propensity of firms to collaborate in innovation projects.

As pointed out in the preceding section, firms wishing to launch an innovation strategy are faced with various alternatives; each with its own costs and risks. The first option is to undertake in-house R&D. In this way, the firm accepts the risk of investing in a specific asset with an uncertain outcome. It also faces the costs of financing a project for which investors will demand guarantees with respect to the final outcome, and justification for the increased costs of supporting a larger innovation structure. The second option is to acquire the necessary technology and results from outside, possibly by outsourcing the research work. The firm has to face the costs incurred in the search for opportunities and assessment of the various options, as well as the costs of potential opportunism on the part of sales representatives. ICT adoption could reduce the costs involved in the make-or-buy decision.

The third possibility open to firms is to collaborate in a joint R&D project with other enterprises. Some of the costs associated with this third alternative coincide with those of in-house R&D and others with those of the *market* alternative. They include some of the so-called transaction costs resulting from the search for the right partners, potential post-contractual opportunism between partners, negotiation of the potential profit share out, the creation of an appropriate framework for collaborative innovation and, of course, information processing and communication. The new ICT help to reduce these costs, minimise risk, and increase the appeal of collaborative R&D. In this context, researchers and analysts have stressed the importance of opportunities for firms to increase and enrich their market connections by managing transactions directly through electronic commerce (Evans & Wurster, 2000; Kalakota, Oliva, & Donath, 1999). Electronic networks have considerably lowered transaction costs by reducing information asymmetries, thus increasing online customisable content in the electronic market (Hawkins, Mansell, & Steinmueller, 1999; Kaplan & Sawhney, 2000).

ICT adoption in firms generally appears to promote inter-firm collaboration (Salas, 2001). Collaboration with suppliers by outsourcing, for example, is enhanced by technology such as the Internet (Porter, 2001), while collaboration with firms in other sectors can help to provide customers with a fuller service (Wielemaker, Elfring, & Volberda, 2001).

One of the most consistent and robust findings in the sociology literature concerns the positive effect of communication on collaboration (Kollock, 1998). The Information Technologies will enhance cooperative innovation by speeding up the transmission of explicit knowledge (Nonaka & Konno, 1998). By definition, explicit knowledge is that which can be conveyed to documents and easily transmitted from one person to another. Some ICT, such as e-mail or teleconferencing, will facilitate the transmission of documents and communication between researchers. Though innovation is largely based on tacit knowledge, which is much more difficult to transmit via ICT (Hislop, 2002), it will always include a component of codifiable knowledge that is easy to transmit.

The use of ICT in collaborative research will save both time and money: tools such as e-mail, teleconferencing, multimedia channels, and so forth reduce unnecessary travelling and the need for trivial discussion. E-mail, meanwhile, enhances relationships between research groups as well as reducing the risk of personal conflicts and culture clash; provides a solution to possible language problems; and ultimately facilitates the transmission of knowledge. The use of this kind of technology is helping to narrow the distance and increase interdependence between geographically distant locations (Salas, 2001). The adoption of this kind of technology will therefore help to establish proper channels of communication between partners, thus increasing their chances of success in any collaborative innovation project.

Any kind of collaboration, and this includes collaborative innovation, reduces duplication and overlap and fosters learning between partners. It must, however, be undertaken with full awareness of the geographical distance separating the partners who may belong to different cultures and even speak different languages. This requires coordination and control, tasks of tremendous importance, which are greatly facilitated by means of ICT (Finnegan & Longaigh, 2002).

Furthermore, the use of some ICT, such as the Internet, helps to create a more fluid relationship with the environment and facilitates firms' access to information of any kind, thus increasing their knowledge base. Through the Internet, firms are able to discover other organisations working on the same focus area and find a way to become part of the cooperation network.

All these reasons lead us to believe that the adoption of information and communications technology will support inter-firm collaboration.

Nevertheless, though it is true to say that the ICT improves research and knowledge diffusion and thus boosts collaborative innovation, it is probably equally true to say that different means of communication have different effects (Jensen, Farnham, Drucker, & Kollock, 2002). Thus, local network application, for example, will be aimed more at increasing cohesion within the firm's in-house research groups. By using the technology inherent to its internal research group, the firm will be able to create its own knowledge base, and thus enhance its absorptive capacity and improve its potential to collaborate with other firms (Cohen & Levinthal, 1990). Kamel and Davison (1998) report that, in addition to the cost and time savings mentioned earlier, ICT will improve the group's decision-making quality, by reducing negative pressure that can lead to problems, such as individual members being overwhelmed by the majority opinion. Tools such as the Internet, e-mail, or inter-firm networks, meanwhile, will help to increase cooperation between firms.

One example of the influence of ICT in collaborative innovation is the Plantania case (Engelbart, 2001; International Service for National Agricultural Research [ISNAR], 2003). This case study looks at the Plantania project, where three companies worked to establish an international value-added and market-oriented chain for potted plants. Floriculture in Europe has traditionally been strongly product-oriented: the growers, produce in bulk, what goes to the auctions, and customers take what is offered. However, with an increasing choice of floricultural products, consumers are demanding better assortment: year-round supply of top-quality produce at a reasonable price. To meet these demands, floricultural agro-chains need to become market oriented. The goal of Plantania was to develop and implement an international value-added and market-oriented chain for potted plants, by improving information flows and setting up a chain quality assurance system. The actors (companies, research institutes, and third parties)[5] were aware of the need to switch from a product orientation to a consumer orientation and of the need for higher quality standards and compliance with emerging consumer demand that products and production be environmentally friendly. Development of forward contracting and logistical improvements were also needed. For this, new information and communication technologies could be used to ease communication within the chain and improve its performance and potted-plant sales.

The Classification of ICT and Their Impact on Collaborative Innovation

In order to understand the full impact of network technologies on collaborative innovation, we will now take a closer look at the different kinds of ICT (Chiarvesio, Di Maria, & Micelli, 2004):

Enterprise computing solutions, that is: applications that support codified information management and processes, such as enterprise resource planning (ERP). Information technologies, such as the ERP system, improve firm efficiency through high-quality data management related to structured and codified processes. (Chiavesio et al., 2004)

Computer-supported cooperative work (CSCW), that is, solutions aimed at facilitating the coordination and collaboration of people within and between organisations (i.e., groupware). These electronic solutions improve the effective management of work distributed among different locations. Internal networks will strengthen the group's internal cohesion and facilitate the creation of a knowledge base. These internal networks will, in addition, enable firm managers to deliver, to their research staff, information of interest relating to innovation, such as calls to participate in innovation projects with other public or private organisations. The firm's personnel will be able to announce the results of their ongoing research activities and undertake any kind of initiative to enable them to improve their own knowledge base and promote collaboration with others. Internal networks of this nature are in operation in many organisations such as the Public University of Navarra in Spain.

Furthermore, inter-organisational networks will facilitate rapid information exchange with customers, suppliers, and other organisations and thus the distribution of work and collaboration in innovation projects. This type of ICT facilitate more frequent, more rapid, and therefore more efficient, interaction with other organisations. In this way they are better able to meet customer needs and undertake innovations in production processes such as the use of new materials and so forth, with organisations close to them in the value chain.

Web-based applications such as multimedia tools, and the Internet. Expectations, particularly those relating to a reduction in transaction costs, are based on more efficient information processes, carried out via the Internet, enabling firms to enjoy the significant benefits resulting from e-commerce (Kalacota et al., 1999; Kaplan & Sawhney, 2000). Internet and e-mail are particularly useful in the early stages of collaboration. E-mail will facilitate the initial contact between different organisations. The Internet enables firms to identify organisations involved in areas related to their own, alerts them to firms seeking partners for cooperation, informs them of administrative measures that promote collaborative innovation, provides them with details of business networks with which to collaborate, and so forth. An example of this is the European Union's Internet service, Community Research & Development Information Service (CORDIS), which helps firms seeking partners for collaborative innovation projects. This service also includes a number of European support programs for collaborative innovation projects, and information on participating firms' completed or ongoing projects. It also provides information on the achievements made in the various projects together with the names of the participating firms. EUREKA is a similar European initiative. Their database contains information on programmes, projects, partner-seeking firms, and so forth.

The Internet also provides an excellent showcase for the firm to publicise its current activities and principal achievements, in order to build up its reputation and promote itself as a potential partner for other firms. Such is the case of University College Dublin, which uses its Web site Innovation and Technology Transfer Centre to offer its services for the promotion of contract research and other forms of university-industry cooperation. It also gives details of its research experience, which in this case is ICT-focused.

CSCW and Web-based applications, have the clearest impact on collaboration in innovation projects. This may be because they simplify the search for partners and facilitate rapid information exchange, or because they can be used to create a knowledge base that will lead to greater cohesion in the working group.

ICT in the European Framework

In addition to their impact on the decision to collaborate in innovation, as shown in the previous section, ICT are widely agreed to have a major impact on a country's productivity and economic growth. The level of ICT use varies greatly across the countries of the developed world. Table 2 shows 2002, 2003, and 2004 expenditures on Information Technology and Communication, not only in the European Union (15), but in the United States and Japan.

The data in Table 2 show that investment in Information Technology (IT) in the European Union as a percentage of gross domestic product (GDP), is lower than in Japan and well below the United States level. Moreover, over the last three years it is possible to observe a worrying downward trend, increasing the gap that separates us from the United States and Japan. The biggest European investors in IT are Sweden and the United Kingdom, who are close to the United States' expenditure level, while Greece and Spain trail behind the rest of Europe in this respect.

The picture changes when we switch to the analysis of Communication Expenditure. In this case, European Union (15) standards are similar to those of the United States, though not as close to those of Japan. Portugal, Sweden, and Greece are the top spenders when it comes to Communication (as a percentage of GDP), with France and Denmark at the bottom end of the ranking.

Table 2. Information technology and communication expenditures as a percentage of GDP (Eurostat)

	Information Technology Expenditure			Communication Expenditure		
	2002	2003	2004	2002	2003	2004
European Union (15)	3.2	3.1	3.0	3.4	3.3	3.3
Belgium	3.1	2.9	2.9	3.6	3.5	3.5
Denmark	3.6	3.5	3.5	3.2	3.2	3.2
Germany	3.1	3.0	3.1	3.0	3.0	3.1
Greece	1.4	1.3	1.3	4.2	4.0	3.8
Spain	1.8	1.8	1.7	3.8	3.6	3.5
France	3.5	3.3	3.3	2.7	2.6	2.7
Ireland	2.3	2.0	2.1	3.6	3.3	3.3
Italy	2.1	2.0	1.9	3.3	3.3	3.4
Netherlands	4.1	3.8	3.8	3.7	3.6	3.7
Austria	3.1	3.0	3.0	3.4	3.4	3.4
Portugal	2.2	2.1	2.0	5.0	5.0	5.1
Finland	3.7	3.6	3.7	3.4	3.4	3.4
Sweden	4.9	4.5	4.4	4.3	4.3	4.3
United Kingdom	4.7	4.2	4.2	3.9	3.8	3.7
Switzerland	4.5	4.2	4.2	3.4	3.4	3.5
United States	4.7	4.6	4.6	3.4	3.3	3.2
Japan	3.5	3.5	3.6	4.3	4.3	4.4

Development levels in the different kinds of ICT technologies both in the European Union, as a whole, and individual countries of Europe for 2003 and 2004 are shown in Table 3. These data, drawn from Eurosat, refer to some of the Web-based applications described earlier; variables such as percentage of enterprises having access to the Internet (e-iacc) and percentage of enterprises having a Web site/homepage (e-web). It also captures variables that we would class as CSCW such as percentage of enterprises using extranet/intranet (e-extra, e-intra). The last of the variables included is a measure of share of Internet sales (e-comm).

As Table 3 shows, the percentage of enterprises having access to the Internet (e-iacc) is very high in all European countries, with values around the 90% mark. When we turn to data for the percentage of enterprises having Web site/homepage (e-web), however, we find wider cross-border discrepancies. Sweden (82%), Finland (75%), and Denmark (81%) take the lead, versus other countries, such as Spain (44%), where very few firms have a Web site. There is every indication that the Internet in some countries (such as Spain) is used more for information search and communication by e-mail than as a means for researchers to publish articles and firms to announce their current activities, needs, and so forth. In the last year, however, there has been a significant increase in the number of firms with a Web site/homepage. Meanwhile, it can be seen that hardly any use is made of e-commerce (share of Internet sales) by European firms to develop their business.

Table 3. Percentage of enterprises having access to the Internet (e-iacc). Percentage of enterprises having Web site/homepage (e-web). Percentage of enterprises using extranet/intranet (e-extra, e-intra). Share of Internet sales (e-comm). (Eurostat)

	e-iacc		e-web		e-extra		e-intra		e-comm	
	2003	2004	2003	2004	2003	2004	2003	2004	2003	2004
EU (25 countries)	:	89	:	58	:	12	:	33	:	2.1
EU (15 countries)	86	90	54	60	12	14	28	34	1.1	2.2
Belgium	91	96	62	68	21	23	41	45	1.8	1.5
Denmark	97	97	75	81	16	16	30	33	1.6	4.4
Germany	95	94	71	72	:	15	22	39	0.7	2.7
Greece	88	87	52	49	10	5	36	32	0.2	1.0
Spain	82	87	33	40	12	17	29	33	0.3	0.4
France	83	:	26	:	14	:	38	:	:	:
Ireland	86	92	59	59	13	12	34	38	10.0	12.8
Italy	83	87	47	44	10	14	28	30	0.3	1.2:
Luxembourg	85	90	58	59	23	:	44	:	:	:
Netherlands	86	88	61	66	28	14	28	28	:	:
Austria	89	94	66	71	:	3	:	26	0.9	1.0
Portugal	70	77	25	29	16	21	30	35	0.6	1.3
Finland	97	97	70	75	18	18	34	36	:	:
Sweden	95	96	80	82	17	15	43	42	2.8	:
United Kingdom	80	87	63	66	6	7	26	30	1.5	2.3

Finally, a variable that could be used to measure the role of ICT in facilitating coordination and collaboration between individuals, both between firms and in-house, is the percentage of enterprises using extranet/intranet (e-extra, e-intra). The data show that intranets are much more widely used than extranets, both at the European Union and the country level. This appears to suggest that the new technologies are used more to promote internal cohesion and coordination than collaboration between organisations.

In the developed world, national ICT strategies are beginning to integrate policies to promote the use of this technology to boost economic growth and employment. This is helping to strengthen the ties between economic development agencies and technology agencies and extra efforts are being made to coordinate policies in this area in order to maximise their impact.[6]

The Empirical Analysis

This section presents an empirical analysis of the impact of a number of issues reported in the literature and the influence of the new information and communication technology on collaborative innovation.

The study is based on data relating to Spanish manufacturing firms, collected by the Spanish National Institute of Statistics in their 1998 and 1996 surveys[7] on technological innovation in firms. The unit of analysis is the firm, and the data is obtained through a questionnaire that was distributed among the sample firms.[8] We focus our attention on innovating firms during the period 1996-1998, while possessing a database on 1,649 firms surveyed in the two waves of the survey, 1994-1996 and 1996-1998. Of these firms, 760 (46.1%) took part in joint R&D and innovation projects with other organisations during the period 1996-1998, while 889 (53.9%) did not.

To test the effect of the new ICT and the aforementioned characteristics on the firm's likelihood to collaborate in innovation, we use a logit model in which the dependent variable is cooperation in innovation. This is a dependent binary variable that takes a value of 1 if the firm has cooperated in R&D and innovation during the period 1996-1998 and 0 otherwise. In the 1998 questionnaire, cooperation in innovation is defined as active participation in joint R&D and innovation projects with other institutions and those of the firm's own projects that are officially linked to other institutions.[9] Mere outsourcing, in which the firm plays no active part, is not considered collaboration. Because of the characteristics of the dependent variable, which is binary, the logit model was chosen to test the impact of some variables on the propensity to collaborate in innovation.

Before proceeding, we will give a brief overview of logit models (Maddala, 1983). These are models based on the assumption that there exists an underlying response variable y_i* defined by the regression relationship

$$y_i^* = \beta' x_i + u_i \qquad (1)$$

In practice, y_i* is unobservable. What is observed is a dummy variable, defined by

$$y=1 \text{ if } y_i^*>0 \qquad (2)$$
$$y=0 \text{ otherwise}$$

In this formulation, βx_i is not $E(y_i/x_i)$, as in the linear probability model, but $E(y_i^*/x_i)$. From the two relationships above it follows that

$$\text{Prob}(y_i=1)=\text{Prob}(u_i>\beta' x_i)=1-F(-\beta' x_i) \qquad (3)$$

where F is the cumulative distribution function of u.

In this case the observed values of y are the result of a binomial process, in which the probabilities are given in (3) and will vary as a function of x_i. The likelihood function, therefore, is

$$L=\prod_{y_i=0} F(-\beta' x_i) \prod_{y_i=1} (1-F(-\beta' x_i)) \qquad (4)$$

The functional form of F will depend on the hypothesis that is made on the random term in the initial expression. If the cumulative distribution of u is the logistic distribution, we have the logit model. In this case

$$F(-\beta' x_i) = \frac{\exp(-\beta x_i)}{1+\exp(-\beta x_i)} = \frac{1}{1+\exp(\beta x_i)} \qquad (5)$$

Therefore,

$$1\text{-}F(-\beta' x_i)=\frac{\exp(\beta x_i)}{1+\exp(\beta x_i)} \quad (6)$$

In our logit model, the independent variables are as follows.

To take firm size into account, we create three dummy variables for three size categories: large for firms with over 250 employees, medium for firms with between 50 and 249, and small for those with less than 50.

To measure technological intensity in firms, we construct three variables: high intensity, medium intensity, and low intensity that take a value of 1 if the firm belongs to a sector in any of the above intensity categories and 0 otherwise. The data supplied by the Institute of Statistics includes the branch of activity in which the firm is involved. The various industrial activities were assigned to the high, medium, or low technological intensity categories on the basis of the OECD classification.[10]

Several variables are used to measure firms' R&D capacity. Systematic I+D takes a value of 1 if the firm replies affirmatively to the question of whether its innovation activity throughout the period 1996 to 1998 included systematic internal R&D and 0 otherwise.[11] This variable allows us to capture whether or not the firm performs internal R&D on a systematic basis. A firm's in-house R&D activities are considered to be continuous or systematic when the firm in question has each year employed at least one full-time researcher and plans to continue in the same way. Otherwise, R&D activities are considered to be sporadic. Since it takes time to develop in-house R&D capacity, we consider a firm to have achieved it if it described its R&D activities as systematic in both questionnaires. We give an overall assessment of non-systematic to any firms that claim to have performed systematic R&D in 1996 and to have reverted to sporadic R&D activity in 1998. The same applies to firms declaring sporadic R&D in 1996 and systematic R&D in 1998.

Two variables are used to capture R&D capacity based on external activities. Thus, technology acquisition 96 takes a value of 0 when the firm declares that it has acquired no new technology during the period 1994-1996 and 1 when it claims to have acquired some. Finally, cooperation in R&D 96 takes a value of 1 if the firm has cooperated in R&D during the 1994-1996 period and 0 otherwise.

A set of binary variables is used to reflect firms' utilisation of the range of information and communications technology available up to year-end, 1998. Among these, there is a group formed by the technologies that we have labelled CSCW. This includes local technical network, which takes a value of 1 if the firm uses a local area network computer for technical data and 0 otherwise; local

Table 4. Descriptive statistics: Mann-Whitney U Test

	Full sample	Collaborators	Non-collaborators	Z
Large	47.54%	56.97%	39.48%	-7.087**
Medium	37.23%	32.50%	41.28%	-3.676**
Small	15.22%	10.53%	19.24%	-4.906**
High inten.	31.59%	34.34%	29.25%	-2.218*
Medium inten.	22.26%	23.42%	21.26%	-1.051
Low inten.	46.15%	42.24%	49.49%	-2.946**
Syst. R&D	51.08%	64.60%	40.83%	-9.627**
Techn. Acq. 96	57.49%	63.82%	52.08%	-4.803**
Cooper.96	37.54%	56.84%	21.03%	-14.964**
Loc. tech. N'wk	65.31%	71.05%	60.40%	-4.527**
Loc. Fy N'wk.	61.55%	66.84%	57.03%	-4.081**
Interfirm N'wk	26.99%	32.89%	21.93%	-4.996**
Internet/e-mail	77.62%	82.63%	73.34%	-4.511**
N	1649	760	889	

Significance level: ** $p<0.01$; * $p<0.05$

factory network, which takes a value of 1 if the firm uses a local area network computer for factory data and 0 otherwise; and inter-firm network, which takes a value of 1 if the firm uses a computer network to connect the factory to subcontractors, suppliers and/or customers and 0 otherwise. The first two of these variables foster knowledge development and transmission within the firm, while the last, inter-firm network, facilitates knowledge transmission *between* firms. Finally, Internet/e-mail is a variable belonging to the third of the groups mentioned earlier, Web-based applications. This takes a value of 1 if the firm uses this communication tool and 0 otherwise.

Before estimating the logit model, we ran a non-parametric Mann-Whitney U test, to see whether the two resulting sub-samples are similarly distributed with respect to the independent variables, when taking the dependent variable cooperation in innovation between 1996 and 1998 as a grouping variable. The results are shown in Table 4.

As can be observed, the degree of up-take of the various information and communications technologies by innovating firms varies with the type of technology. Thus, web-based applications, such as the Internet and e-mail, are the most widely used; 77.62% use them. CSCW, meanwhile, varies according to whether the variables refer to within-firm collaboration, as in the use of local network computers for technical data (65.31%) and factory data (61.55%), or inter-firm collaboration, as when there is a network connecting the firm to its suppliers or customers. This last option is taken up by only 26.99% of the sample firms. This sample of firms can be seen to exhibit a similar performance pattern to that observed in the various European countries.

If we compare the situation in firms where innovation is collaborative and those where it is non-collaborative, we can see that there is a higher level of ICT adoption among collaborating firms than among the rest. This is in fact the most significant difference between the two groups.

The data displayed in Table 4 show that the collaborative firm profile is one of large firms belonging to highly technology-intensive sectors. They also tend to have engaged on a systematic basis in in-house R&D. In addition, a higher percentage of these firms made technology acquisitions and collaborated in R&D between 1994 and 1996.[12]

The logit methodology will allow us to test whether the aforementioned characteristics do, in fact, increase the likelihood of cooperating in innovation. Table 5 presents the model estimates. Two models were estimated in order to gain a better understanding of the repercussion of ICT usage in collaborative innovation. The variables included in the first model are size, R&D capacity, and technological intensity, that is, those reported in the literature to have an obvious impact on collaborative innovation. In addition to these, the second model includes, ICT variables, such as local technical network, local factory network, inter-firm network and Internet/e-mail. For the variables in each of the models, the table shows the coefficients, significance levels, Nagelkerke's R-square value, and the percentage of correctly predicted cases. As shown, the models are practically identical in R^2 value or predictive power. This suggests that ICT adoption by firms has a limited impact on collaborative innovation.

The results indicate that size has an impact on the propensity of firms to collaborate in innovation, since large firms are found to be more likely to do so. This result is consistent with findings reported in other studies.[13] The degree of technological intensity in the operating sector, however, has no significant effect on their propensity in this respect. These results are similar in both of the estimated models.

Table 5. Estimates for collaborative innovation

	Model 1	Model 2
Large	0.633***	0.534***
Medium	0.275	0.228
High inten.	0.142	0.117
Medium inten.	0.120	0.066
Syst. R&D	0.362***	0.331***
Tech. Acq. 96	0.128	0.127
Cooper.96	1.334***	1.336***
Loc. tech. N'wk		0.162
Loc. Fy N'wk.		-0.137
Interfirm N'wk		0.241*
Internet/e-mail		0.217
Constant	-1.403***	-1.559***
Nagelkerke's R^2	0.198	0.205
% predicted	68.7	68.7

Significance levels: *** p<0.01; ** p<0.05 *p<0.1

As observation shows, the firm's R&D capacity has a positive and significant effect on the propensity to collaborate in innovation. Thus, systematic in-house R&D throughout the whole period and collaboration in R&D between 1994-1996 are positively and significantly associated with the propensity to collaborate in the 1996-1998 period. It emerges, on the one hand, that in-house research is complemented with outside research in the same area, through collaborative activities. It can be seen, furthermore, that collaboration flourishes better when the firm has past experience in the matter. This is an issue that is widely discussed in the literature on collaboration. However, the external acquisition of technology in the 1994-1996 period does not appear to have any decisive impact on the firm's level of R&D activities. Taken together, these results uphold the absorptive capacity theory that appears in Cohen and Levinthal (1989, 1990) and other studies (Arora & Gambardella, 1994; Bayona et al., 2001; Colombo & Garrone, 1998; Fritsch & Lukas, 2001).

From model 2, in Table 5, the one that includes the ICT variables, it is possible to check that the coefficients and significance levels of the variables obtained in model 1 remain constant. This is an indication of the consistency of the results. As far as the ICT variables are considered, we can see that the one with a significant and positive coefficient is Inter-firm network, which shows that the presence of a computer network connecting the firm with subcontractors, suppliers, and/or customers increases the firms' propensity towards collaborative innovation. This suggests that technologies that serve to strengthen relationships, and perhaps also confidence and commitment between different organisations, are the ones that influence the decision to collaborate in innovation. We are unable to judge from the available data whether the purpose of establishing a network is to facilitate collaboration, or is intended to cover other needs, as might be assumed. This confirms the advanced intuition in the theoretical framework. The remaining CSCW variables considered to help within-group cohesion are not significant. Nor is the *Internet/e-mail* variable, since it is linked to the reduction of transaction costs associated to the search for information and suitable partners. This may be due to widespread use of these tools among all the sample firms.

The influence of the aforementioned variables in collaborative innovation may vary as a function of two firm characteristics: size and technological intensity. Tables 6 and 7 show the results obtained.

Table 6 shows the results when the sample is segmented into large, medium, and small firms. It can be seen that in larger firms (large and medium sized) continuous internal R&D activity has a positive effect on inter-firm collaboration, with both internal and external research taking place simultaneously. Thus, the absorptive capacity theory of Cohen and Levinthal (1989, 1990) is confirmed in the two upper size categories but not in the smaller one. Large firms are also the only ones in which technology acquisition in previous years affects the propensity

Table 6. Estimates for cooperation in innovation: Sorted by size

	Large	Medium	Small
High inten.	0.062	0.734	0.627*
Medium inten.	0.034	-0.055	0.623
Syst. R&D	0.528***	0.371**	-0.525
Tech. Acq. 96	0.296*	-0.122	0.260
Cooper.96	1.205***	1.262***	2.224***
Loc. tech. N'wk	-0.060	0.360	0.119
Loc. Fy N'wk.	0.001	-0.233	-0.103
Interfirm N'wk	0.294*	0.109	0.744
Internet/e-mail	0.203	0.366	-0.075
Constant	-1.119***	-1.309***	-1.791***
N	784	614	251
Nagelkerke's R^2	0.191	0.146	0.236
% predicted	67.0	67.8	76.5

Significance levels: *** $p<0.01$; ** $p<0.05$ *$p<0.1$

to engage in inter-firm collaboration in innovation. As found for the sample as a whole, previous cooperation (in 1996) increases the propensity towards collaboration, with the effect increasing as firm size decreases.

Variables relating to ICT usage practically lose their significance when the sample is segmented by size. The only one to retain significance is inter-firm network in the large firm model.

We finished our analysis by dividing the sample firms into three groups according to technological intensity (high, medium, and low). The results are displayed in Table 7.

The first finding that emerges is that firms with a low to medium level of technological intensity are the ones where systematic in-house R&D has a positive effect on the decision to collaborate in innovation. The high-level technological intensity group is therefore the only one in the whole analysis in which systematic internal R&D activity does not increase the propensity towards collaboration. This is almost certainly due to the fact that R&D is practised by all firms in technology-intense sectors, this characteristic is therefore not a driver of collaborative innovation in this particular segment.

The second result is that the use of a certain type of ICT—the CSCW variables—has an influence on collaborative innovation in technologically intense firms. However, while inter-firm network, which fosters coordination and cooperation between different organisations, is positively associated with collaborative innovation, the local factory network variable presents a negative relationship. Again, this somewhat surprising result is found in technologically intense firms. Research in such firms is routine practice, though it may be more complex and more difficult to transmit. Therefore, if the firm has a local network

Table 7. Estimates for cooperation in innovation: Sorted by technological intensity

	High intensity	Medium intens.	Low intensity
High inten.	0.432	0.380	0.707**
Medium inten.	0.234	-0.042	0.395
Syst. R&D	0.284	0.475**	0.313*
Tech. Acq. 96	0.208	-0.174	0.180
Cooper.96	1.526***	1.164***	1.303***
Loc. tech. N'wk	0.471	-0.051	0.087
Loc. Fy N'wk.	-0.654**	0.258	0.018
Interfirm N'wk	0.688***	-0.098	0.146
Internet/e-mail	0.126	0.617*	0.140
Constant	-1.415***	-1.477***	-1.691***
N	521	367	761
Nagelkerke's R^2	0.261	0.189	0.181
% predicted	71.4	68.4	68.5

Significance levels: *** p<0.01; ** p<0.05 *p<0.1

to facilitate the transmission of information within the group, its researchers are more able to interrelate and communicate with each other, and have less need to collaborate with outside organisations. Collaboration with outsiders will be promoted, however, if there is a network connecting the firm with its subcontractors, suppliers, and/or customers.

The Internet/e-mail variable is significant only when firms belong to a sector of medium technological intensity. This may be because use of the Internet and e-mail is routine in firms of all types, whatever their technological intensity

Conclusions and Challenges for the Future

Certain firm characteristics play a role in the propensity towards collaboration in technology and innovation. The existing literature includes analyses of the way in which the decision to cooperate in R&D is influenced by the firm's size, R&D capacity, and complexity of technological development. Following the development of the absorptive capacity theory by Cohen and Levinthal (1989, 1990), it has been suggested that firms should accompany their external knowledge acquisition with extra R&D. R&D capacity is therefore one firm characteristic that is positively related to collaborative innovation. Several studies find firm size to be another, in so far as larger firms will find it easier to undertake in-house R&D and by doing so increase their capacity to acquire knowledge from potential partners. There are also indications from other perspectives, such as the transaction costs theory, the organisational theory, or the absorptive capacity theory, that the greater the technological complexity of the sector, the more collaborative innovation will take place.

The aim of this chapter, however, is to take a step towards a deeper understanding of collaborative innovation. Hence, we have attempted to estimate to what extent it is influenced by the adoption of the new ICT and by other factors cited in the existing literature. Present-day economics need to be understood at the global level; thus, both ICT and inter-firm collaboration are key issues in firm strategy. This is indeed how they are interpreted by administrations at all levels, including the European Union.

We believe the distinction we make between the different types of ICT to have been worthwhile. Our theoretical reasoning is that tools for increasing within-group cohesion, instruments to facilitate the transfer of results, and knowledge between different organisations' CSCW, access to the Internet, and the use of e-mail will all contribute to increase firms' propensity towards collaborative innovation.

The results of the empirical analysis with a sample of Spanish manufacturing firms help to shed some light on this issue, which has received scant attention in the literature. First, we find that the practice of in-house R&D increases the propensity towards collaborative innovation, in our sample, as does any previous experience in the matter. This appears to confirm the complementarity of internal and external R&D, in addition to a *past experience effect*, which has a bearing on the propensity to collaborate. Secondly, we find confirmation of the impact of firm size, shown by the fact that larger firms are the most active in collaborative innovation. Third, the technological intensity of the operating sector does not appear to have much bearing on the decision to collaborate. Finally, with respect to the ICT variables, from these results it can be concluded that whenever there is a tool to facilitate communication between the firm and its customers and/or suppliers and subcontractors, thus enabling all kinds of knowledge transmission between organisations, the propensity towards collaborative innovation will increase.

From these results it is possible to draw some implications for business managers. One potential strategy for firms wishing to promote collaboration in an attempt to ensure the success of the complex innovation process would be to begin by investing in R&D, in order to improve their innovative capacity. They can then use ICT to enable them to create networks to bring their firm into contact with customers, and/or suppliers, and subcontractors.

Despite the preliminary effort made in this study, this issue still requires deeper investigation. The questions to be answered by future research might include the following: What influence do other types of ICT have on collaborative innovation? What type of knowledge is transmitted through the networks? How far do knowledge transmission networks enhance coordination between research groups, reduce their transaction costs, increase interdependence between different research groups, and encourage the use of emerging communication channels?

Taking a closer look at two firm characteristics: size and technological intensity, our empirical analysis indicates that the impact of ICT utilisation on collaborative R&D is strongest in large firms and technologically intense sectors. Again, results point to a specific type of ICT; the type labelled in this chapter as CSCW, that is, technology designed to facilitate both internal and inter-organisational group coordination.

In light of these results, and bearing in mind the concern shown by the various administrations with respect to the promotion of both ICT and collaborative innovation, as mentioned earlier, we are able to reach the following conclusion. To help drive cooperative efforts to innovate, encouragement and support should be provided for any initiatives aimed at setting up inter-organisational networks to facilitate knowledge transmission between research personnel in different firms and thus ensure that research projects are carried through successfully.

References

Arora, A., & Gambardella, A. (1994). Evaluating technological information and utilizing it. *Journal of Economic Behavior and Organization, 24,* 91-114.

Bayona, C., García-Marco, T., & Huerta, E. (2001). Firms' motivations for cooperative R&D: An empirical analysis of Spanish firms. *Research Policy, 30,* 1289-1307.

Cassiman, B. (1999). Cooperación en investigación y desarrollo. Evidencia para la empresa manufacturera española. (Cooperation on research and development. The case of Spanish industrial firms.) *Papeles de Economía española, 81*, 143-154.

Cassiman, B., & Veugelers, R. (2002). R&D cooperation and spillovers: Some empirical evidence. *American Economic Review, 92*(2), 1169-1184.

Chiarvesio, M., Di Maria, E., & Micelli, S. (2004). From Local networks of SMEs to virtual districts? Evidence from recent trends in Italy. *Research Policy, 33,* 1509-1528.

Coase, R. (1937). The nature of the firm. *Economica, 4*, 386-405.

Cohen, W., & Levinthal, D. (1989). Innovation and learning: The two faces of R&D. *The Economic Journal, 99,* 569-596.

Cohen, W., & Levinthal, D. (1990). Absorptive capacity: A new perspective on learning and innovation. *Administrative Science Quarterly, 35,* 129-152.

Colombo, M., & Garrone, P. (1998). A simultaneous equations model of technological agreements and infra-mural R&D. In M. Colombo (Ed.), *The changing boundaries of the firm* (pp. 140-157). London: Routledge.

Duysters, G., Kok, G., & Vaandrager, M. (1999). Crafting successful strategic partnerships. *R&D Management, 29,* 343-351.

Engelbart, F. (2001). Plantania: A business case description. 'S-Hertogenbosch: Agri-Chain Competence Center (ACC). Retrieved August 5, 2005, from http://www.kc-acc.org/pdf/plantania.pdf

European Commission. (2000). Nuevo marco para la investigación europea. *Innovación & Transferencia de Tecnología, 3,* 4-5. (New framework for the European. *Innovation and Technology Transfer, 3*, 4-5.)

European Commission. (2001). Llegan las vacas gordas? *Innovación & Transferencia de Tecnología, 2,* 5-6. (Are the boom years coming? *Innovation and Technology Transfer, 2*, 5-6.)

Evans, P., & Wurster, P, (2000). *Blown to bits: How the new economics of information transform the strategy.* Boston, MA: Harvard Business School Press.

Finnegan, P., & Longaigh, S. (2002). Examining the effects of information technology on control and coordination relationships: An exploratory study in subsidiaries of pan-national corporations. *Journal of Information Technology, 17,* 149-163.

Fritsch, M., & Lukas, R. (2001). Who cooperates on R&D? *Research Policy, 30,* 297-312.

Girma, S. (2002, August). *Absorptive capacity and productivity spillovers from DFI: A threshold regression analysis.* Working Paper. University of Nottingham, Leverhulme Centre for Research on Globalisation and Economic Policy.

Granstrand, O., Patel, P., & Pavit, K. (1997, Summer). Multi-technology corporations: Why they have distributed rather than distinctive core competencies. *California Management Review, 39*(4), 8-25.

Griffith, R., Redding, S., & Van Reenen, J. (2000). *Mapping the two faces of R&D: Productivity growth in a panel of OECD industries.* Discussion Paper, nº 2457. London: CEPR.

Hagedoorn, J. (1996). Trends and patterns in strategic technology partnering since the early seventies. *Review of Industrial Organization, 11,* 601-616.

Hagedoorn, J., & Schakenraad, J. (1994). The effect of strategic technology alliances on company performance. *Strategic Management Journal, 15,* 291-309.

Hamel, G. (1991). Competition for competence and inter-partner learning within international strategic alliances. *Strategic Management Journal, 12,* 83-103.

Haüsler, J., Hohn, H. W., & Lütz, S. (1994). Contingencies of innovative networks: A case study of successful interfirm R&D cooperation. *Research Policy, 23*, 47-66.

Hawkins, R., Mansell, R., & Steinmueller, W. E. (1999). Toward digital intermediation in the information society. *Journal of Economic Issues, 33*(2), 383-391.

Hislop, D. (2002). Mission impossible? Communicating and sharing knowledge via information technology. *Journal of Information Technology, 17,* 165-177.

Innobarometer. (2004). *Flash Eurobarometer 164.* EOS Gallup Europe.

International Service for National Agricultural Research (ISNAR). (2003). *Estudio de caso Plantania.* (Case study: Plantania.) Retrieved August 2, 2005, from http://isnar.cgiar.org/shiip/es/plantania.htm

Jensen, C., Farnham, S., Drucker, S., & Kollock, P. (2002). The effect of communication modality on cooperation in online environments. In *Proceedings of CHI 2000,* The Hague, The Netherlands.

Kalakota, R., Oliva, R. A., & Donath, B. (1999). More over, e-commerce. *Marketing Management, 8*(3), 22-32.

Kamel, N., & Davison, R. (1998). Applying CSCW technology to overcome traditional barriers in group interactions. *Information & Management, 34,* 209-219.

Kaplan, S., & Sawhney, M. (2000) E-Hubs: The new B2B marketplaces. *Harvard Business Review, 78*(3), 97-103.

Kay, J. A. (1993). Fundamentos del éxito empresarial. Editorial Ariel, Barcelona. (*Foundations of corporate success*. Oxford University Press.)

Kelly, K. (1998). *New rules for the new economy: 10 ways the network economy is changing everything.* London: Fourth Estate.

Kleinknecht, A., & Van Reijnen, J. (1992). Why do firms cooperate on R&D: An empirical study. *Research Policy, 41,* 551-560.

Kogut, B. (1988). Joint ventures: Theoretical and empirical perspectives. *Strategic Management Journal, 9,* 319-332.

Kogut, B., & Zander, U. (1992). Knowledge of the firm, combinative capabilities, and the replication of technology. *Organization Science, 3*(3), 383-397.

Kollock, P. (1998). Social dilemmas: The anatomy of cooperation. *Annual Review of Sociology, 24,* 183-214.

Maddala, G. S. (1983). *Limited-dependent and qualitative variables in econometrics.* Cambridge, MA: Cambridge University Press.

Malone, T. W., Yates, J. A., & Benjamin, R. (1987). Electronic markets and electronic hierarchies. *Communication of the ACM, 30*(6), 487-497.

Narula, R. (2001). *R&D collaboration by SMEs: New opportunities and limitations in the face of globalisation.* Mimeo: University of Oslo and the Norwegian School of Management.

Narula, R., & Hagedoorn, J. (1999). Innovating through strategic alliances: Moving towards international partnerships and contractual agreements. *Technovation, 19,* 283-294.

Nelson, R. R., & Winter, S. G. (1982). *An evolutionary theory of economics change.* Cambridge: Harvard University Press.

Nonaka, I., & Konno, N. (1998). The concept of Ba: Building a foundation for knowledge creation. *California Management Review, 40*(3), 40-54.

OECD (2002). The measurement of scientific and technological activities. *Oslo Manual.* París.

OECD (2004b). *OECD Information Technology Outlook.* (2004 ed.). Paris: OECD Publications.

Patel, P., & Pavitt, K. (1994). *Technological competencies in the world's largest firms: Characteristics, constraints and scope for managerial choice.* Discussion Paper, nº 13. SPRU, STEEP.

Polanyi, M. (1966). *The tacit dimension.* London: Reutledge and Kegan Paul.

Porter, M. (2001, March). Strategy and Internet. *Harvard Business Review,* 63-78.

Robertson, T., & Gatignon, H. (1998). Technology development mode: A transaction cost conceptualization. *Strategic Management Journal, 19,* 515-531.

Rocha, F. (1997). *Inter-firm technological cooperation: Effects of absorptive capacity, firm-size and specialization.* Discussion Series Paper, nº 9707, The United Nations University (UNU) Intech, Maastrich, The Netherlands.

Salas, V. (2001, November 10-11). La empresa española en el umbral de la sociedad de la información Ponencia presentada en el IV Congreso de Economía de Navarra, Pamplona, Spain. (The Spanish firm on the threshold of information society. In *Proceedings of the Fourth Conference of Economy of Navarre.*)

Santangelo, G. (2000). Corporate strategic technological partnerships in the European information and communications technology industry. *Research Policy, 29,* 1015-1031.

Shing, K. (1997). The impact of technological complexity and interfirm cooperation on business survival. *Academy of Management Journal, 40*(2), 339-367.

Steensma, H. K. (1996). Acquiring technological competencies through inter-organizational collaboration: An organizational learning perspective. *Journal of Engineering and Technology Management, 12,* 267-286.

Teece, D. (1986). Profiting from technological innovation: Implications for integration, collaboration, licensing and public policy. *Research Policy, 15,* 285-305.

Teece, D. (1992). Competition, cooperation and innovation. Organizational arrangements for regimes of rapid technological progress. *Journal of Economic Behavior and Organization, 18,* 1-25.

Teece, D. (1996). Firm organization, industrial structure, and technological innovation. *Journal of Economic Behavior and Organization, 31,* 193-224.

Tidd, J., & Trewhella, M. (1997). Organizational and technological antecedents for knowledge creation and learning. *R&D Management, 27,* 359-375.

Tyre, M., & Hauptman, O. (1992). Effectiveness of organizational responses to technological change in the production process. *Organization Science, 3*(3), 301-320.

Wielemaker, M., Elfring, T., & Volberda, H. (2001). How well-established firms prepare for the new economy. *International Studies of management and Organization, 31,* 7-29.

Williamson, O. (1975). *Markets and hierarchies: Analysis and antitrust implications.* The Free Press.

Winter, S. (1987). Knowledge and competence as strategic assets. In D. J. Teece (Ed.), *The competitive challenge: Strategies for industrial innovation and renewal* (pp. 159-184). New York: Harper and Row.

Endnotes

The authors acknowledge support from Spanish National Institute of Statistics for helping with data collection. Financial support from MCYT grants n°. SEJ2004-07242-C03-02 and SEC2003-06457 and the government of Navarre is gratefully acknowledged.

1 The term transaction is a synonym for the transfer from one agent to another of the right to use certain goods. It is the contractual phase that accompanies the release/cession libramiento of goods and services in exchange operations, which are essential to the functioning of an economic system in which there is private property and division/allocation of labour. Transaction costs can be defined as all the costs incurred in carrying out transactions. These are basically access to information and potential opportunism on the part of the agents involved. Transaction costs are information costs (search costs to identify potential partners), negotiation

costs, (costs of negotiating and drawing up the contract) and guarantee costs (monitoring and supervising and enforcing the contract).

2 Of all the available surveys, this is the only one with a questionnaire in which the Spanish National Institute of Statistics includes items relating to the level of ICT adoption and participation in cooperative innovation projects.

3 It is the nature of new knowledge that basically determines its ease of imitation. An innovation or new technology may incorporate either tacit or codifiable knowledge (Polanyi, 1966; Winter, 1987). Codifiable or explicit knowledge is that which can be easily transmitted and received. It is vulnerable to industrial espionage and, unless it is patented or registered, will be easy to imitate. Thus, the competitive advantage achieved by it will not be sustainable. Codifiable knowledge can be conveyed in plans or computer diskettes and is easily interpretable. Some knowledge, however, is tacit; that is, difficult to express and therefore to transfer. It resides in employee skills and in the organisational routines by which they are coordinated (Nelson & Winter, 1982). For tacit knowledge to be imitated, therefore, the same routines and abilities must be reproduced in other firms; an objective that is not easily achieved.

4 Following Tyre and Hauptman (1992) technological complexity refers to the novelty or sophistication of new equipment and its technological characteristics. For Steensma (1996), complexity is inherent to a technology and is beyond the firm's control. Complex technology is highly sophisticated and difficult to pass on to the user.

5 Consumer-oriented organisation, ICT company, consultancy firm, among others.

6 A table with the priority areas of ICT policy in the OECD countries is included in the Appendix (Table 8).

7 The study focuses on the reply to two surveys conducted by the Spanish National Institute of Statistics. The reason for using 1998 data 1998 is that it is the only year in which the survey included questions relating to firms' adoption of certain technologies, including ICT. Thus, it is possible to examine the relationship of these technologies to collaborative R&D. Moreover, in the theoretical framework, mention was made of the potential importance of a firm's ICT capacity in collaborative innovation. This capacity cannot be achieved in a single period; it is the result of an accumulation of knowledge generated through innovation activities over some time. These activities may be in-house R&D, technology acquisition or the participation in joint research projects with other organisations. Information from the 1996 survey indicating whether the firms have undertaken any of these activities is used in conjunction with 1998 data to enable us to obtain a closer approximation of the firms' innovation capacity.

[8] Though the Spanish National Institute of Statistics does not give the survey reply rate, it is an official body and, therefore, firms are obliged under pain of fine to respond.

[9] The survey does not include information on the nature of the projects (basic or applied research), how many the firm is involved in, the amount of funding allocated to each, the contractual form used in the agreements, the duration of projects, and so forth.

[10] Shown in Table 9 of the Appendix.

[11] Both the 1996 and 1998 questionnaires ask firms whether their R&D activities are systematic or sporadic.

[12] Differences are significant between the two sub-samples on all these variables.

[13] See the works of Hagedoorn and Schakenraad (1994); Häusler et al. (1994); Cassiman and Veugelers (1998); Colombo and Garrone (1998); and Fritsch and Lukas (2001).

Appendix

Table 8. ICT priority areas in OECD countries (OECD, 2004)

ICT priority areas
General policies
ICT policy environment
Fostering ICT innovation
Research and development programs
Government development projects
Increasing diffusion and use
Professional/managerial ICT skills
Government on-line. Government as model users
Diffusion to businesses
Diffusion to individuals and households
ICT marketing environment
Competition in ICT markets
Intellectual property rights
Enhancing the infrastructure
Broad band
Electronic settlement/payment
Standards
Promoting trust on-line
Security of information systems and networks
Assessment and evaluation

Table 9. Industries with a high, medium or low technological intensity following an adaptation of the OECD classification

Industries
High technological intensity
Chemicals and pharmaceutical products
Office machinery and computer equipment
Electrical machinery and material
Electronic components
Radio and TV apparatus and communications
Medical, precision and optical instruments
Medium technological intensity
Rubber and plastic materials
Mechanical machinery and equipment
Motor vehicles
Marine construction
Aeronautic and space construction
Other transport material
Low technological intensity
Extraction, cooking plants and petrol refining
Food, drinks and tobacco
Textiles, clothing and footwear
Timber and cork (excluding furniture)
Furniture
Other manufacturing activities
Recycling
Paper, publishing, graphic arts and reproduction
Non-metallic mineral products
Ferrous and non-ferrous metal products
Metal products (excluding machinery and equipment)
Electricity, gas and water production and distribution

Chapter VII

Inter-Organizational Information Systems and Strategic Alliances: Symbiosis or Competition?

Sebastián Bruque-Cámara, University of Jaén, Spain

José Moyano-Fuentes, University of Jaén, Spain

Abstract

In this chapter we analyze the relationships between the structures of inter-organizational information systems (IOS) and strategic alliances. The relationships between both structures tends to be framed within a collaborative dynamic in which the IOS serves the objectives of the alliance, thereby reinforcing the links that bind the alliance members together. However, in this chapter we also suggest that there may be a non-collaborative—that is, competitive—relationship between the IOS and the strategic alliance, especially when both structures compete for firms from the same sector and geographical area. To test our hypotheses we have analyzed a sample of 162 firms belonging to the agro-food sector in Spain.

Introduction

Information and communications technology (ICT) and information systems (IS) have had—and indeed continue to have—growing influence on the way firms are managed and organized. There can be no doubt that the lion's share of the modern organization's value chain has been changed in one way or the other by the introduction of computing, robotics, or telecommunications tools (Porter & Millar, 1986). ICT is likely to have a major impact on the components of the value chain that are most closely related to supplies, transport, or data processing. Moreover, in recent decades most of the primary or support activities of firms' value chains have gone through a process of intensification in terms of their need for information management, so ICT has been playing an increasingly important role within organizations.

However, one component of the ICT revolution that we have experienced (and that is still going on) has not taken place within the limits of the traditional organization, but has originated in the relationships that link organizations with each other. The sharing of resources and the collaboration (at the same time as competition) between organizations has meant that a large part of firms' managerial, financial, and productive resources have begun to acquire a virtual dimension. By means of this virtualization firms can make use of resources and capabilities that are geographically dispersed and, more importantly, that originally belonged to other institutions. The technological revolution therefore leads to a revolution in the relationships between firms and organizations, which put resources and capabilities into common use with the aim of improving individual competitive positions. The main advantage of inter-organizational cooperation comes not only from the possibility of accessing virtualized resources but of doing so much more cheaply than would be the case if these resources were acquired on the market for their individual use.

A direct consequence of this virtualization of resources and of the cooperation between institutions is the appearance of IOS (Kauffman, 1966; Konsynski, 1993) and of strategic alliances and networks (Gulati, Nohria, & Zaheer, 2000). Strategic alliances and IOS have traditionally been regarded as two sides of the same coin. Thus, many strategic alliances and stable cooperation agreements have as their main aim to create or strengthen an IOS (Medina Garrido & Bruque Cámara, 2004; Pigni, Ravarini, Sciuto, Zanaboni, & Burn, 2004). In this way, firms can either reduce the uncertainty inherent in today's complex environments or alternatively cut the transaction costs deriving from their relationships with other organizations (Ciborra, 1990; Gulati et al., 2000; Malone, Yates, & Benjamin, 1987).

However, the literature has shown little interest in situations in which the IOS and the strategic alliance interact competitively. In these cases the appearance of an IOS could actually destabilize the system of psychological or economic incentives that bind the members of a strategic alliance together. The disparity of structures and interests, or simply mutual ignorance, can lead to situations in which the IOS erodes the structure of the alliance. Thus, in this chapter we attempt to gain more knowledge about this question, illustrating the situation that occurs when IOS and alliance act competitively—that is, non-symbiotically—and suggesting some explanations for the problem. These explanations will provide us with the basis for offering a number of implications and recommendations, using data from the olive oil production sector in Spain.

This chapter is organized into five sections following on from this present introduction. In the second section we review the state of the question and propose our working hypotheses. In the third we describe the methods utilized to test the hypotheses. In the fourth we outline our findings. In the fifth section we outline our main conclusions, while in the final section we offer a series of recommendations for firms and public administrations, as well as directions for future research.

Background and Hypotheses

The relationship between the use of IOS and opting for the market (autonomy), hierarchy, or strategic alliance, has been widely studied in the literature (Brynjolfsson, Malone, Gurbaxani, & Kambil, 1994). In most previous analyses researchers have demonstrated that the IOS and strategic alliance come into being in a coordinated way with the aim of achieving similar objectives (Henderson & Venkatraman, 1993). The IOS serves to achieve the objectives set down by the alliance in terms of information management and the conversion of this information into knowledge. In parallel, the strategic alliance gives the IOS legitimacy in the eyes of the alliance members, as well as financial and human resources (Ciborra, 1993; Gurbaxani & Whang, 1991; Malone et al., 1987). However, less attention has been paid to the competitive relationship that may develop when a strategic alliance and an IOS interact in the same competitive sphere (or sector). Neither have we found any work comparing the use of the IOS by firms that are not members of strategic alliances and firms that are. These questions may have important implications for the management of these alliances as well as for the evolution of the IOS.

In a strategic alliance the functions of search, selection, analysis, and distribution of information are carried out by the structure of the alliance itself. Strategic alliances, when they take the form of a stable strategic network (Miles & Snow, 1986), tend to develop an ad hoc governance structure (Grandori & Soda, 1995). The governance structure takes responsibility for managing the activities within the remit of the alliance (joint commercialization, research and development, supply chain management, joint supplies, etc.). This ad hoc structure, which is a consequence of the alliance, tends to take the form of a new organization (joint venture, new capital firm, or horizontal cooperative arrangement) (Grandori & Soda, 1995; Gulati, 1998) charged with developing the elements (technological tools, personnel, and departments) required to carry out the activities for which the alliance is responsible.

On the other hand, IOS have been seen as useful tools in firms' management. There is substantial evidence of the various advantages firms can obtain from cooperating in an IOS. The IOS is a cooperation structure in itself; it aims to achieve goals such as reducing operational and management costs, sharing information that is strategic from the competitive perspective, reducing negotiation or information costs in economic transactions, strengthening the links that bind strategic alliance members, developing knowledge management systems that prove useful to all IOS participants, modifying the market structures in order to increase IOS participants' bargaining power, and so forth (Bakos, 1991; Eom, 2005; Kumar & Van Dissel, 1996).

When the information provided by an IOS lies within the decision scope of the alliance, the interface between the IOS and alliance members will be the alliance ad hoc governance structure. In this respect, according to Thorelli (1986), the alliance as a whole can be regarded as a single organization with defined strategic objectives. The ad hoc structure responsible for dealing with questions entrusted to the alliance should obtain, analyze, handle and if necessary, disseminate the information coming from the IOS. Thus, we can say that the information coming from the IOS will be distributed among the firms making up the alliance, respecting the decision and information flows previously established between the ad hoc governance structure and the alliance members. In other words, there should not be any direct use by the firms making up the alliance (individually regarded) of the resources provided by the IOS. In the case of non-alliance members, however, it is the firms themselves that must directly obtain the information and resources they need to carry out their activity. If the IOS proves to be an effective tool, firms outside the alliance will try to access and use the system in order to improve their competitive position in the sector or simply to survive in a complex and dynamic environment. In short, under the conditions discussed previously, firms not belonging to a strategic alliance are likely to use the IOS more than firms forming part of a strategic alliance. This idea is reflected in Hypothesis H1:

Hypothesis H1: Firms not forming part of a strategic alliance use IOS more than strategic alliance members.

The previous hypothesis allows us to infer that firms forming part of a strategic alliance—individually considered—will have less need to access and use the IOS, since the ad hoc governance structure created by the alliance takes over the functions of obtaining and analyzing information. This ad hoc structure will be responsible for accessing and analyzing the information, and if necessary, distributing it to the alliance members respecting the traditional flows of decision and communication between the ad hoc structure and the member firms. In this case we can say that the IOS does not interfere either with the internal organization or with the decision-making structure operating within the alliance. In other words, there is a concordant relationship between the structure of the IOS and the structure of the alliance.

However, the internal situation of the strategic alliance will not always allow this concordant relationship to arise. The intensity with which firms become involved in their strategic alliances can vary depending on a number of explanatory factors (Medina Garrido, 2003; Thorelli, 1986). Firms in alliances are all formally members of the same structure, but some may behave opportunistically (Gulati et al., 2000), exploiting some of the advantages provided by the alliance on the one hand, and on the other, making key decisions independently and considering only their own individual interests. In these situations alliance members' degree of involvement with the alliance, measured by their delegation of decision making to the governance structure, will decline (Arino & de la Torre, 1998).

Opportunistic behavior appears to be associated with alliance members' reduced involvement with respect to the principles of cooperation and also with respect to the joint operation of the strategic alliance (Álvarez, Barney, & Bosse, 2004). When opportunistic behavior occurs, firms' individualistic orientation in the decision-making process means that they have a greater need to access key information. One of the ways of individually accessing key information for decision making is to make use of IOS (Porter & Millar, 1986). Thus, we would suggest that the reduced involvement that results from opportunistic behavior should lead to greater direct use of the IOS by firms behaving opportunistically. In these cases the flow of decisions and information established between the alliance governance structure and the alliance members will weaken since direct flows between the IOS and some alliance members will emerge; in other words, a discordant relationship develops between the structure of the IOS and the structure of the strategic alliance. In short, under the premise of opportunistic behavior we would expect members to become less involved in their alliance, making them more likely to make direct use of the IOS. In the context of reduced involvement and direct use of the IOS, discordant relationships would emerge between the IOS and strategic alliance structures. This idea is reflected in Hypothesis H2.

Hypothesis H2: Within a strategic alliance, the reduced involvement of member firms with their alliance is associated with an increased direct use of IOS.

Methods

Data

To test the previous hypotheses we analyzed a sample of firms belonging to the olive oil production sector in the province of Jaén (Spain). This province produces 45% of the total Spanish production of olive oil. Spain, in turn, produces 40% of world production, according to the International Olive Oil Council (Consejo Oleícola Internacional, 2001; Moyano-Fuentes & Núñez-Nickel, 2004a). In 1998 the population of olive oil producing firms in the province of Jaén amounted to 323 organizations (Moyano-Fuentes & Núñez-Nickel, 2004b).

The data collection procedure was based on a questionnaire administered either by telephone or in a face-to-face interview. For this, the research team first contacted the firms selected in order to explain to them the objectives of the research. If the interlocutor was appropriate (chief executive or general manager) and if they agreed to participate in the study, we proceeded to carry out the survey by telephone. In some cases the managers requested that they be able to respond to the questionnaire in a face-to-face interview, so we then arranged for a subsequent interview in which the respondent could complete the questionnaire.

We used random sampling, stratified in proportion to the legal status of the firm (cooperative/non-cooperative)[1] and geographic area (district). We introduced the first stratification criterion in order to keep the same proportion of cooperative firms in the sample as in the global population. This precaution was necessary because cooperatives are more prone to participate in strategic alliances than non-cooperative firms (Torres-Ruiz, 1998). The stratification criterion by geographic areas, in turn, aimed to ensure that responses from urban areas were not over-represented to the detriment of rural areas. The final number of valid responses was 162, which represents 50.15% of the total population. Of the 162 firms participating, 94 habitually use IOS and 61 belong to stable strategic alliances in the form of horizontal cooperative arrangements.[2] These strategic alliances are made up of five organizations on average, and they started to appear in the sector from 1980 onward. The remaining technical specifications of the sample are reported in Appendix I.

As far as the IOS is concerned, we used as reference in this study, the Poolred System, set up by the Olive and Olive Oil Promotion and Development Foundation, which began to operate in January 1997. We chose this system because it is widely used among the firms in the sector and has ample experience in obtaining, analyzing, handling, and disseminating information generated by the markets for bulk olive oil in Spain. At present, the system is used by 541 institutions (producers, public administrations, and bottling firms) in Spain, and in 2003 it handled a volume of information relating to transactions valued at over 1.3 billion euro.

Questionnaire

The questionnaire for the current research contained a total of 39 questions, in which we included open questions, Likert-type closed questions and closed questions of the semantic differential-type. The Likert-type and semantic differential-type scales each consisted of five intervals. This chapter is part of a wider research project, so we have not used all the items included in the questionnaire here. Specifically, we only used the data pertaining to the use of IOS, membership of strategic alliances, and degree of involvement in the alliance.

Variables

In this section we describe the variables used in the research, as well as how they are measured.

Participation in strategic alliances. Dichotomous variable taking value 1 if the firm belongs to a stable strategic alliance (horizontal cooperative arrangement) and 0 if it does not belong to such an alliance.

Use of IOS. Dichotomous variable taking value 1 if the firm uses the Poolred System[3] and 0 otherwise.

Involvement in strategic alliance. These variables were measured using five five-point Likert-type scales. These scales measure the intensity with which the member firm uses the alliance mechanisms in its decision making, especially with decisions having important implications for the performance or survival of the organization. The items are as follows: (1) products are commercialized exclusively through the strategic alliance; (2) there are periodic contacts with the alliance's governance structures for consultation on economic and management aspects; (3) relationships with suppliers are channeled through the alliance's governance structure; (4) the remaining members in the alliance are consulted about economic and management aspects; and (5) relationships with the public administrations are channeled through the alliance's governance structure. In all cases 1 corresponds to never and 5 very frequently.

Analysis

Table 1 shows the means, standard deviations, and correlations between the variables used.

To test the proposed hypotheses we opted to carry out an analysis based on descriptive statistics and non-parametric statistical tests to observe the differences between members and non-members of strategic alliances. Specifically, we used contingency tables and the non-parametric Mann-Whitney U test, to test Hypothesis H1 (see Tables 2 and 3), and descriptive statistics and the Mann-Whitney U test, to test Hypothesis H2 (see Tables 4 and 5).

Table 1. Descriptive statistics and Spearman correlations

Variable	Mean (sd)	1	2	3	4	5	6
1. Participation in strategic alliances	0.38 (0.49)						
2. Use of IOS	0.42 (0.50)	-0.14					
3. Commercialization through alliance	2.72 (1.50)	--	-0.22				
4. Consultations with governing structure of alliance	3.33 (1.84)	--	-0.08	0.61**			
5. Relationships with suppliers through alliance	2.37 (1.25)	--	-0.02	0.53**	0.43**		
6. Consultations with other alliance members	1.75 (1.16)	--	-0.30*	0.50**	0.37**	0.40**	
7. Relationships with public administrations through alliance	1.89 (1.45)	--	0.03	0.47**	0.32**	0.40**	0.41**
*Correlation significant at 0.05. **Correlation significant at 0.01.							

Table 2. Test of hypothesis H1. Contingency table

		Participation in strategic alliances (%)		Total
		NO	YES	
Use of IOS	NO	53 (53)	41 (67)	94
	YES	47 (47)	20 (33)	67
	Total	100 (100)	61 (100)	161

Table 3. Test of hypothesis H1. Mann-Whitney U test

	Participation in strategic alliances	No. of firms	Mean rank	Sum of ranks	Mann-Whitney U	Asymptotic sig. (bilateral)
Use of IOS	NO	100	85.33	8533.50	2616.5	0.077†
	YES	61	73.89	4507.50		
† Significant difference at 90%.						

Results

The results of the analyses carried out here allow us to draw a series of conclusions, which we now outline. With regards to Hypothesis H1, the descriptive statistics appearing in the contingency table (Table 2) support the relationship postulated in Hypothesis H1, although quite weakly. Observing the percentages in Table 2, the results show that 47% of the sample firms that do not participate in alliances use the IOS, while only 33% of alliance members directly use the system. However, the Mann-Whitney U test does not allow us to generalize this result and say that in the population as a whole firms not participating in alliances use IOS more than alliance members, since the differences observed between the two groups have a very weak statistical significance. In short, although the data collected in the sample are coherent with Hypothesis H1, it is only possible to assure that this situation occurs in the global population at the 90% level.

Hypothesis H2 is also only weakly supported by the statistical tests carried out here. At first sight (Table 4) the descriptive statistics appear to support the postulates of Hypothesis H2. The means corresponding to each involvement variable, ordered by groups (in function of whether they use the IOS or not) result in higher values when the firms do not use the IOS than when they do. Only once

Table 4. Hypothesis H2: Descriptive statistics of groups[a]

	Use of IOS	No. of firms	Mean	s.d.
3. Commercialization through alliance	NO	38	3.45	1.77
	YES	17	3.06	2.01
4. Consultations with governing structure of alliance	NO	38	2.92	1.40
	YES	16	2.25	1.65
5. Relationships with suppliers through alliance	NO	37	1.76	1.26
	YES	16	2.19	1.83
6. Consultations with other alliance members	NO	35	2.40	1.19
	YES	16	2.31	1.40
7. Relationships with public administrations through alliance	NO	38	1.89	1.13
	YES	17	1.41	1.18

[a]Calculations only include firms forming part of strategic alliances.

Table 5. Test of hypothesis H2. Mann-Whitney U test

	Use of IOS	No. of firms	Mean rank	Sum of ranks	Mann-Whitney U	Asymptotic sig. (bilateral)
3. Commercialization through alliance	NO	38	28.78	1093.50	293.50	0.561
	YES	17	26.26	446.50		
4. Consultations with governing structure of alliance	NO	38	29.62	1093.50	223.50	0.116
	YES	16	22.47	446.50		
5. Relationships with suppliers through alliance	NO	37	26.69	987.50	284.50	0.791
	YES	16	27.72	443.50		
6. Consultations with other alliance members	NO	35	26.17	916.00	274.00	0.897
	YES	16	25.63	410.00		
7. Relationships with public administrations through alliance	NO	38	30.84	1172.00	215.00	0.024*
	YES	17	21.65	368.00		

* Significant difference at 95%.

does this not occur—with the variable measuring relationships with suppliers through the alliance (variable 5). Considered overall, the data collected in Table 4 allows us to indicate that for the sample firms the inverse association relationship referred to in Hypothesis H2 is satisfied. That is, and again for the sample, firms belonging to alliances but not very involved in them use the IOS more than the firms that are more involved.

However, the results referred to in the previous paragraph cannot be generalized to the whole population, since the Mann-Whitney U tests do not result in statistically significant differences (this only occurs for the variable measuring involvement through relationships with the public administrations). In short, although the data from the sample is in line with Hypothesis H2, we cannot generalize them to the population as a whole.

Summary

This work is one of the first attempts to link the degree of involvement of alliance members with the extent to which they use IOS. Although the results are not entirely satisfactory from the perspective of statistical significance or the possibility of generalization, they undoubtedly represent a starting point from which we can advance our understanding of the interrelations—sometimes interferences—that can develop between different cooperation structures that coexist in the organizational framework (the same sector). These structures (on the one hand strategic alliances, and on the other IOS) can be used symbiotically, ensuring that one structure reinforces the other. However, as we have been able to observe in this chapter, when the two cooperation structures (IOS or alliance) are set up at different times and backed by different institutions, a negative (or competitive) effect may be induced, which ends up weakening the bonds that exist within the strategic alliance.

The appearance of an IOS can weaken the structure of the alliance in two different but related ways. First, it can undermine some of the motivations encouraging firms to form strategic alliances in the first place. One of these motivations has to do with the access to more and better information, allowing firms to reduce uncertainty in dynamic and complex environments (Ciborra, 1993). If the firm can obtain this information in an IOS that remains outside the alliance it will be less likely to strengthen its links with the alliance. The second motivation concerns the opportunistic behavior of some alliance members, who will feel freer to act independently if they have access to key information autonomously. Taking this situation to the extreme, the disparity between the IOS and the strategic alliance may lead to the breakup of the latter, or even its disappearance. Whether the alliance fails, however, will depend on the degree of rivalry between the governance structures of the alliance and the IOS as well as on the strategic importance of the information managed by the IOS.

In short, this chapter attempts to provide new, useful explanations for the situations in which strategic alliances find themselves weakened by the appearance of cooperation structures (IOS) that erode part of their own responsibilities. Our findings may therefore prove interesting for the research stream analyzing the factors leading to the failure of cooperation between firms, as well as provide appropriate solutions (Harrigan, 1985; Kale, Dyer, & Singh, 2002; Mitchell & Singh, 1996). However, the work may also have implications for managers, particularly at the following three decision levels: the managers of the governance structure of the IOS, the managers of the strategic alliance's governance structure, and public administrations interested in promoting stable cooperation agreements. These management implications, together with possible lines of research for the future and the limitations of the work, are analyzed in the final section.

Recommendations and Future Trends

From the perspective of the management of strategic alliances, managers of these structures should not only be conscious of the opportunities but also the threats represented by the appearance of an IOS within the scope of activity of the alliance. This threat may be greater when the information provided by the IOS is closely linked to the activity controlled by the alliance. In these cases, alliance members can be motivated to behave opportunistically, or even simply to abandon the cooperation structure. The threat of the IOS is also greater when it is set up totally independent of the strategic alliance, that is, when it is backed by institutions outside the alliance, which may even be rivals.

We now offer a series of recommendations that can help to mitigate the situation that has been previously described. In all these limitations we have to take into account that for creating a strong and stable alliance it is necessary to provide value to the members of the alliance. There are several ways to create value such as improving logistics, decreasing production or transaction costs, improving the access to new markets, improving the organization's bargaining power, and so forth. For the case of interactions between IOS and strategic alliances we can provide some recommendations that may prove useful to increase the value transferred by the IOS to the strategic alliance. First, since the threat derives from the difference in the interests and structures between both systems, it is necessary to initiate a process of negotiation between the governance structures of the two systems (alliance and IOS). Thus, there is a need for a process of macro-cooperation between both structures, with the intention of aligning the interests of both systems. The optimal option for the alliance and the least harmful for the IOS would be for a single link between the alliance and the IOS to be formed by the alliance's governance structure itself. In exchange, the strategic alliance could help to strengthen the IOS, providing it with new resources or helping it to obtain, process, and disseminate information. In other words, this strategy of macro-cooperation would have the ultimate aim of establishing a symbiotic relationship between both structures and ensuring that neither interferes negatively with the other. We should point out that in this case the scope of the macro-cooperation could be limited to aspects relating to the strategic implications of the information handled by the IOS, so that both structures could maintain their autonomy in all the remaining areas.

The managers of the IOS governance structure could also benefit from the process of negotiation started with the strategic alliance. In this case, the IOS should weigh up the advantages of the aforementioned strategy of macro-cooperation: the strengthening of the structure itself; greater financial, technical, and human resources; and the broadening of the scope of the IOS itself.

Third, public administrations could play a fundamental role in the process of macro-cooperation initiated between both cooperation structures. In situations of rivalry or divergence of interests the public administrations could intervene to reestablish the balance of power, arbitrate compensation to one or another structure, or simply bring the managers of both cooperation structures into contact in order to facilitate negotiation. The function of the public administration is even more relevant when the sector in which the IOS and alliance operate can be regarded as strategic for the public interest, or when collaboration between the IOS and alliance could serve as a useful example for other key sectors.

With regards to advancing research, we might make some recommendations for the academic and research community. First, it is necessary to further the knowledge of the factors that might provoke the weakening or failure of inter-organizational cooperation structures (alliances or IOS). Greater understanding of the causes of success or failure may have positive implications, since it could inform public and private policies aimed at preventing situations of risk from arising. Second, there is a need to deepen understanding of the factors that encourage firms to become more involved in the various formulas of inter-organizational cooperation, thereby preempting opportunistic behaviors among the cooperating firms. Third, in the particular case of interference between the alliance and IOS, we would recommend that researchers attempt to generalize the results reported in this current chapter. It would be particularly interesting to see further research in other sectors, or in other countries in which differences of national culture may play an important role. Finally, it would be useful to obtain more information about the function that the structure of the sector (size and number of firms, technological development, transparency of market, etc.) may play in the relationships between strategic alliances and IOS.

As we have already mentioned, this work suffers from a number of limitations, which should be borne in mind when judging the significance of the results obtained here. In the first place, the statistical tests carried out are not entirely satisfactory, in the sense that they do not always allow us to confirm the association relationships postulated in the hypotheses. For this reason some conclusions are only applicable to the sample and not to the population as a whole. Second, in the proposed associations another series of moderating variables may play a role in clarifying the causes and effects that intervene in the problem. Thus, the greater propensity to use IOS of firms with little involvement in their alliances could be moderated by the initiation of corrective measures by the governance structure of the alliance. Other variables may also have intervened in this relationship, such as the culture of the organizations participating in the alliance; the existence of a collaborative relationship between the IOS structure and the governance structure of the alliance; or the degree of technological maturity of the institutions involved (governance structure, member firms, IOS management structure, etc.). Third, the design of the measuring instruments (dichotomous and Likert perceptual variables) may have proved inadequate for revealing complex relationships between the constructs under analysis. Fourth, we have analyzed the relationships between IOS structures and strategic alliances in a very specific sector (olive oil production sector in Spain), so we encourage other researchers to analyze this topic in other sectors. Finally, in this study we only analyze one type of strategic alliances (horizontal alliances) so further research is needed in the case of other types of alliances such as vertical alliances, diagonal alliances, or internal networks (Hinterhuber & Levin, 1994).

References

Álvarez, S. A., Barney, J. B., & Bosse, D. A. (2004). Trust and its alternatives. *Human Resource Management, 42*(4), 393-404.

Arino, A., & de la Torre, J. (1998). Learning from failure: Towards an evolutionary model of collaborative ventures. *Organization Science, 9*(3), 306-325.

Bakos, J. Y. (1991). Information links and electronic marketplaces: The role of inter-organizational information systems in vertical markets. *Journal of Management Information Systems, 8*(2), 31-52.

Barron, D. N. (1995). Credit unions. In G. R. Carroll & M. T. Hannan (Eds.), *Organizations in industry* (pp. 137-162). New York: Oxford University Press.

Brynjolfsson, E., Malone, T. W., Gurbaxani, V., & Kambil, A. (1994). Does information technology leads to smaller firms? *Management Science, 40*(12), 1.628-1.644.

Ciborra, C. (1990). *Technologie di coordinamento.* Milano: Angeli. (Technology for coordination. Milano: Angeli.)

Ciborra, C. (1993). *Teams, markets, and systems: Business innovation and information technology.* Cambridge: Cambridge University Press.

Consejo Oleícola Internacional. (2001, de 2 de abril). *Proyección de la producción y del consumo del aceite de oliva en el horizonte de 2008* (CE/R.54/Doc. Num. 3). Madrid: Consejo Oleícola Internacional. (International Olive-Oil Council. [2001, April 2]. *Olive-oil production and consumption projections on the 2008 horizon* (CE/R.54/Doc. Num. 3). Madrid, Spain: International Olive-Oil Council.)

Eom, S. (2005). *Inter-organizational information systems in the Internet age.* Hershey: Idea Group Publishing.

Grandori, A., & Soda, G. (1995). Inter-firm network; antecedents, mechanisms and forms. *Organization Studies, 16*(2), 183-214.

Gulati, R. (1998). Alliances and networks. *Strategic Management Journal, 19*(4), 293-317.

Gulati, R., Nohria, N., & Zaheer, A. (2000). Strategic networks. *Strategic Management Journal, 21,* 203-215.

Gurbaxani, V., & Whang, S. (1991). The impact of information systems on organizations and markets. *Communications of the ACM, 34*(1), 59-73.

Harrigan, K. (1985). *Managing for joint venture success.* New York: Praeger.

Henderson, J. C., & Venkatraman, N. (1993). Strategic alignment: Leveraging information technology for transforming organizations. *IBM Systems Journal, 32*(1), 4-16.

Hinterhuber, H. H., & Levin, B. M. (1994). Strategic networks—The organization of the future. *Long Range Planning, 27,* 43-53.

Kale, P., Dyer, J. H., & Singh, H. (2002). Alliance capability, stock market response, and long-term alliance success: The role of the alliance function. *Strategic Management Journal, 23,* 747-767.

Kauffmann, F. (1966, January-February). Data systems that cross company boundaries. *Harvard Business Review,* 20-35.

Konsynski, B. R. (1993). Strategic control in the extended enterprise. *IBM Systems Journal, 32*(1), 1-20.

Kumar, K., & Van Dissel, H. G. (1996). Sustainable collaboration: Managing conflict and cooperation in inter-organizational systems. *MIS Quarterly, 20*(3), 279-300.

Malone, T. W., Yates, J., & Benjamin, R. I. (1987). Electronic markets and electronic hierarchies. *Communications of the ACM, 30*(6), 484-497.

Medina Garrido, J. A. (2003). *El papel de las tecnologías de la información en las redes estratégicas*. Tesis Doctoral. Cádiz: Universidad de Cádiz. (*The role of IT on strategic networks*. Doctoral thesis. Cádiz: University of Cádiz.)

Medina Garrido, J. A., & Bruque Cámara, S. (2004). Las alianzas estratégicas entre PYMEs para crear negocios electrónicos. (Strategic alliances among SMEs to promote e-business.) *Revista Futuros, 2*(5), 1-17.

Miles, R. E., & Snow, C. C. (1986). Network organizations: New concepts for new forms. *California Management Review, 28,* 62-73.

Mitchell, W., & Singh, K. (1996). Survival of business using collaborative relationships to commercialize complex goods. *Strategic Management Journal, 17*(3), 169-196.

Moyano-Fuentes, J., & Núñez-Nickel, M. (2004a). El tamaño de la población como determinante de la probabilidad de desaparición organizativa. (Population size as a determinant of the likelihood of organizational failure.) *Revista Europea de Dirección y Economía de la Empresa, 13*(1), 11-24.

Moyano-Fuentes, J., & Núñez-Nickel, M. (2004b). *La supervivencia empresarial en la industria de extracción de aceite de oliva giennense durante el período 1944-1998.* Jaén: Instituto de Estudios Giennenses. (*Organizational survival in the olive-oil production industry in Jaén [Spain] in the period 1944-1988*. Jaén: Instituto de Estudios Giennenses.)

Pigni, F., Ravarini, A., Sciuto, D., Zanaboni, C. A., & Burn, J. (2004). Business associations as hubs of inter-organizational information systems for SMEs—The two cities portal. In S. O. Eom (Ed.), *Inter-organizational information systems in the Internet age* (pp. 170-210). Hershey: Idea Group Publishing.

Porter, M. E., & Millar, V. E. (1986). Cómo obtener ventajas competitivas por medio de la información. (How information technology gives you competitive advantage.) *Harvard Deusto Business Review, 25,* 3-20.

Thorelli, H. B. (1986). Networks: Between markets and hierarchies. *Strategic Management Journal, 7,* 37-51.

Torres-Ruiz, F. J. (1998). *Las cooperativas jiennenses y la comercialización de los aceites de oliva: Una perspectiva estratégica.* Jaén: Instituto de Estudios Giennenses. (*Olive-oil marketing by cooperatives in the province of Jaén, Spain: A strategic approach*. Jaén: Instituto de Estudios Giennenses.)

Endnotes

1 The cooperative firm is constituted as a democratic organization and its management is based on the principle of one member, one vote. The other organizational forms (capitalist forms) have the characteristic that the number of votes is directly linked to the size of the shareholding (Barron, 1995, p. 138).

2 A horizontal cooperative arrangement is an alliance made up of firms that are cooperative firms. The objective of a horizontal cooperative arrangement is to jointly manage some activities of the value chain (such as marketing, logistics, R&D, etc.).

3 The Poolred System is an IOS run by the Foundation for Promotion and Development of Olive Oil. All the firms that are included in the system have to send, online, the data related to their transactions (daily olive oil sales). Transactions data is subsequently organized and presented to the other members of the IOS. The members of the system may access market information through an intranet that is the same intranet they use for introducing their data in the system.

Appendix I. Technical Specifications

Population	Olive oil producing firms in province of Jaén (Spain) (323 organizations).
Time	December 2003-March 2004.
Type of sampling	Probabilistic, stratified in proportion to legal form (cooperative/non-cooperative) and geographic location.
Sampling units	General manager or CEO.
Sample errors	±5.55% for p = q = 0.5 and confidence level 95.5%.
Sample size	162 valid responses.
Research technique	Telephone survey combined with face-to-face survey.

Chapter VIII

The Role of IT in Family Firm Internationalization through Strategic Alliances

Alfonso Miguel Márquez-García, University of Jaén, Spain

Guadalupe Fuentes-Lombardo, University of Jaén, Spain

Sebastián Bruque-Cámara, University of Jaén, Spain

Abstract

In this chapter we analyze the potential impact of information technology (IT) on the internationalization processes of family firms via cooperation agreements. With this in mind, we address the main research questions concerning the management of family firms and internationalization processes via strategic alliances. Subsequently, we carry out an analysis to determine IT's impact on family firms' internationalization processes. Our conclusions and the future trends that we observe suggest that family firms should undertake a series of internal adjustments in order to exploit, with more possibility of success, the changes resulting from the technological integration processes that can be expected in the near future.

Introduction

Until recently the family firm as an object of research did not spark much interest in the scientific community. However, the growing awareness of the enormous importance of these firms in the economic structures of most developed countries[1] both in terms of wealth generation and employment; their role as one of the main agents of the economy at the national and international level (Klein, 2000; Morck & Yeung, 2003; Shanker & Astrachan, 1996); as well as the existence of certain specific idiosyncratic characteristics of these firms that require attention and study to understand their existence; and how they are configured and operated, has led to a growing understanding of the need for research into family firms.[2]

As a consequence, research on family firms has grown significantly in recent years,[3] focusing on an analysis of the differentiating characteristics of these firms, their strengths and weaknesses, and the unique property deriving from the superimposition of the firm management and family systems.[4]

These strengths and weaknesses influence these firms' strategic orientations (Harris, Martínez, & Ward, 1994). Among the various growth strategies available, internationalization is an option that is becoming more important due to the globalization of markets and the convergence in consumers' tastes, both favored by the possibilities offered by the appearance, diffusion, and generalization of information and communication technologies among businesses.

The geographic diversification of family firms' markets takes on different characteristics compared to non-family firms, since the reasons why these firms implement internationalization processes, as well as how they go about it, may be affected by their particular characteristics.

The option to internationalize the firm through strategic alliances tends to be regarded as the best way of initiating a cooperation process, since it can increase the speed of entry into new markets, and the firm has to make lower and less risky investments, cutting the transaction costs and facilitating access to resources and capabilities that would be more difficult to obtain in the market. However, there may be problems of cultural compatibility between the international partners, which are heightened by the family or non-family nature of the firms.

Internationalization is difficult in and of itself, but we should also consider the potential difficulties deriving from choosing cooperation as the way to achieve it, bearing in mind the influence that the family has in the firm—both in the decision to internationalize and in the choice of how to carry this strategy out. In any case, both in the decision to internationalize and to cooperate, IT can play a key role, either by reducing production or transaction costs, or as a medium through which to access valuable and rare resources that are not readily transferable in the market.

In short, this chapter analyzes the role of IT as a facilitating element in the internationalization processes undertaken by family firms via strategic alliances. We examine the advantages and disadvantages of the main forms of internationalization that these firms may employ in their strategy of international expansion. After selecting the most appropriate entry modes for family firms, we analyze the role of IT in each phase of the cooperation process. We end with some comments about the future trends in IT, internationalization, and family firms, which lead to our main conclusions.

Background

Internationalization Strategy in Family Firms

In spite of the importance of family firms, there is still a lack of consensus about the aspects that should be considered to conceptualize this type of firm. Most definitions from the literature are worded to take into account one or several of the following aspects: first, who manages and/or controls the ownership of the capital; second, who has the authority to manage or make the important decisions; third, the family's level of involvement in the business; and fourth, the possibility of a generational handover, or the desire to keep the firm in the hands of the succeeding generations (Handler, 1989; Heck & Scannell, 1999).

According to Fuentes, Vallejo, and Martínez (2005, p. 168) "a family firm is one in which the members of a single family possess a part of the ownership that allows them to exert a significant influence—through their culture—in the management and strategic control of the business, and in which there is also a shared vision of growth and desire to keep the business in the hands of the succeeding generation". As well as containing the four aspects previously mentioned, this definition also considers that one of the objectives of this type of firm should be to seek growth; because, if this is not the case the firm's very survival could be endangered, when the number of family members rises and the corporate wealth fails to keep pace. Thus, family firms need to grow, and one of the possible strategies to achieve growth is to embark on a process of internationalization. In Table 1 summarizes the strengths and weaknesses characterizing family firms. These factors could affect the desire for growth and the choice of internationalization strategy to achieve it.

Most work published up to now on internationalization in family firms analyzes some of the factors that can either favor or limit the geographical expansion of these firms. These factors are a result of the overlap between the family system and the business system and derive from certain aspects of a strategic family or organizational type or other aspects relating to the life cycle of the firm.[6]

Table 1. Strengths and weaknesses of family firms (Fuentes et al. 2005)[5]

Strengths	Weaknesses
Knowledge and experience in the business	Succession or generational handover
Long-term vision	Nepotism
Commitment, dedication, and family pride	Family strife influencing business
Loyalty, trust, and communication	Lack of clear structural definition and overlap of roles
Flexibility and speed in decision making	Financing problems
Convictions, values, and shared vision	Autocratic and paternalistic management style
Customer satisfaction and concern for quality	Feudalization of the organizational structure
Autonomy in running business and financial independence	Moral coercion of descendents

Starting from the concept of a family firm and of its particular characteristics, we consider that these firms' internationalization strategies could have characteristics influencing why and how they geographically expand their markets that differ from those of non-family firms.

For this, we shall review the main theories that have been applied to firms' internationalization strategies. Specifically, we shall consider those theories proposed from the economic approach (Guillén, 2001) and the resource-based view, in order to try to understand the motivations leading firms to consider and embark on geographic expansion of their markets beyond their national territory. We shall then analyze these motivations in the specific case of the family firm.

Within the economic approach, authors propose many motivations for embarking on and developing a strategy of internationalization. Some of these refer to the possession of ownership advantages in certain activities (Dunning, 1979, 1981, 1988, 1992; Hymer, 1976), transferring a firm's experience in manufacturing differentiated products abroad (Caves, 1971), obtaining benefits as a result of internationalization and localization advantages (Buckley & Casson, 1976; Dunning, 1979, 1981, 1988, 1992; Hennart, 1982), or the interest in increasing foreign market commitment (Johanson & Vahlne, 1977, 1990).

Apart from the aforementioned motivations, authors have made various contributions on the firm's internationalization strategy, both theoretical[7] and empirical,[8] under the perspective of the resource-based view. According to this approach, one of the main reasons driving firms to adopt an internationalization strategy is closely related to the possession of certain ownership advantages, as is also argued from the economic approach.

The analysis offered from the resource-based view applied to the internationalization process makes it clear that the firm's set of intangible resources is critical for achieving a sustainable competitive advantage in the target market. Using the classification of this type of resource made by Grant (1991), Hall (1992), and Fernández (1996), we can distinguish between human, technological, organizational, and reputational resources.

In the specific case of family firms, their competitive advantage could be based on the family members' greater knowledge and experience in the business; the result of their long-term orientation and stronger commitment to their firm. This in turn is motivated by the unceasing dedication not only of the present-day family members currently running the business, but very probably of the previous generations as well. In addition, this type of firm appears to be characterized by its greater flexibility, decision-making speed, and customer orientation. Satisfying the customer and seeking quality become two of the main objectives of family firms. Thus, responding to consumers' needs and doing so more quickly requires the presence of certain organizational resources, which this type of firm appears to possess in relative abundance.

Everything suggests that most of the strengths and weaknesses of family firms can be explained by their organizational culture, insofar as they refer to values and virtues that originate precisely in the family nature of the business. Vallejo (2003) concludes that the culture of family firms is characterized by the greater involvement, identification, and loyalty of the employees; the existence of a more favorable organizational climate and higher levels of trust, participation, and cohesion; and by a higher level of reinvestment of the profits. At the same time, he points out that the leadership style of this type of firm generates greater commitment, harmony, and cohesion, and that the long-term oriented management that is practiced positively influences the profitability. These findings lead us to think that the culture of family firms and their management and leadership styles can be converted into fundamental, intangible assets, in view of their positive influence on the firms' internationalization strategy.

In addition, the image of the firm, its reputation, and its brand, can also be regarded as intangible resources upon which its competitive advantage can be built. In the particular case of the family firm, there has always been a special interest in looking after, maintaining, and improving these aspects. However, with regards to the technological resources of these firms, none of the strengths listed in Table 1 lead us to think that they can be considered potential factors of success in these firms' internationalization processes.

In Figure 1 we briefly outline the positive or negative effects that some of the strengths and weaknesses of family firms, related to intangible resources, may have in the internationalization process.

Features of family firms may therefore interfere on internationalization strategies—positively or negatively (Figure 1). However, we should consider that this influence does not have to be similar to the particular case where the family firm

Figure 1. Influence of family firm's strengths/weaknesses in the internationalization process

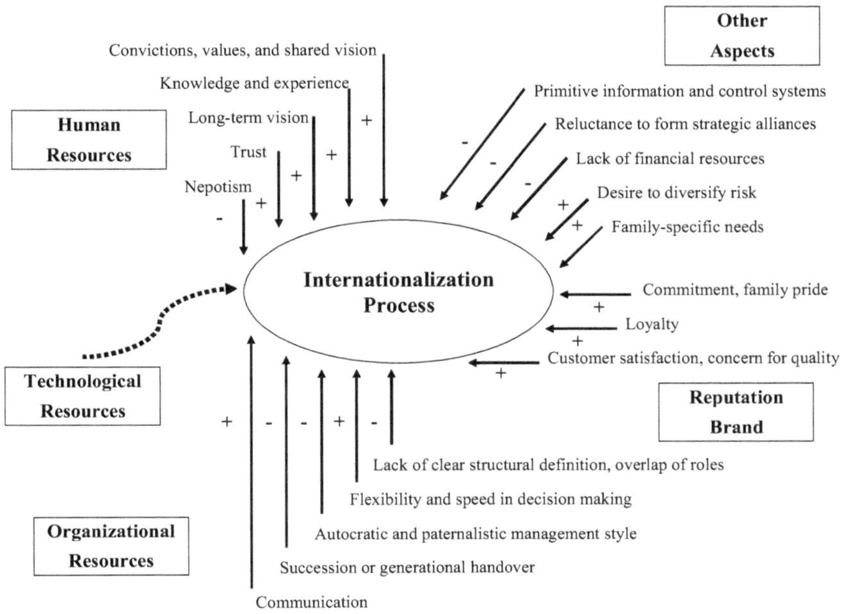

develops strategic alliances in order to get geographical market diversification.[9]

To summarize, when family firms are considering an internationalization strategy, their main motivations may come from their possession of certain intangible resources (human, organizational, reputational, brand, etc.), which will enable them to generate a competitive advantage in the target market. At the same time, the internationalization process could improve a firm's set of intangible resources in its own domestic market. Other motivations could possibly lie in the desire to diversify risks or other aspects relating to certain family needs.

Cooperation as a Mode of Internationalization

Alliances are not a new concept as Ohmae (1989) pointed out, and their use has always been part of the social and political fabric, but until recently, international management theory did not place much emphasis on the study of international business cooperation. In most cases, collaborations were viewed as second-best solutions. Moreover, they were perceived to have a negative impact on competitiveness.[10] However, the attention paid to international cooperation has now increased, in both international management theory and practice (Holtbrügge, 2004, pp. 255-256).

The decision to internationalize the firm through strategic alliances, combining efforts, resources, and talent, tends to be regarded as the best way of initiating this process (DGPYME, 2005; Fernández & Arranz, 1999, p. 207). This is because it not only provides the firm a rapid entry into a new market (Glaister & Buckley, 1996; Hill, Hwang, Kim, 1990), but it also reduces the firm's dependence on resources by cutting the amount it needs to invest. At the same time, the firm takes on less risk, especially if it is relatively inexperienced in foreign markets. Cooperation allows firms to reap economies of scale and scope, improving or consolidating their market position and raising their earnings.[11]

According to Contractor and Lorange (1988, p. 15) it is generally expensive, difficult, and time consuming to create a global organization and achieve a significant international competitive presence. Cooperation can shorten this process of internationalization, facilitating international expansion, and this speed may be critical for obtaining first-mover benefits (Gannon, 1993).

Cooperation in general and international joint ventures in particular have been described (Doz & Hamel, 1998) as the logical response to the intense and rapid changes that are occurring in economic activity, technology, and globalization.[12] Few firms can create and provide products and services alone in this new situation. Thus, international cooperation can provide easier and quicker access to product or process technologies that would otherwise be difficult to obtain individually in the home market (Fernández & Arranz, 1999, p. 57).

The high transaction costs of internationalization operations are thought to be reduced by means of cooperation, especially in situations where there are barriers against other modes of entry (Glaister, Husan, & Buckley, 2004, p. 3-4). It also makes it easier to access resources and capabilities that would be difficult to obtain in the market or acquire individually. Cooperation provides an opportunity for mutual learning[13] (a firm may be highly able to produce a product but be ignorant of the foreign market, so it will depend on its partners), which will facilitate absorption of the local partner's knowledge. However, there may be problems of cultural compatibility between international partners, heightened by the family or non-family nature of the firms.

Transcending the dichotomy between pure exporting and direct investment, cooperation is revealed as a particularly appropriate way of implementing an international strategy, as it opens up a wide range of possibilities allowing firms to combine different levels of risk, resource commitment, and degree of control over the foreign operations (Rubio, 2004, p. 8-9). Thus, according to Dunning (1997), we are currently witnessing a transition of the economic system from market capitalism to alliance capitalism, where firms are linked through different forms of equity and non-equity cooperation.

The transaction cost theory, agency theory, game theory, and the resource-based view for international business cooperation may explain why international strategic partnerships are concluded (Faulkner & de Rond, 2000; Parkhe, 1993; Tsang, 2000), but they do not provide much evidence of how to manage them (Holtbrügge, 2004). With respect to the management structure of international strategic business cooperation, there are basically two distinguishable forms: dominant and shared. While dominant parent structures are characterized by asymmetric power relations, in shared management structures all partners have equal representation in the governance bodies.[14] In this context, we should consider the potential problems caused by the agency relationship that emerges between the parties involved in the cooperation. Stewardship theory could, however, provide elements of analysis of the facilitators of this relationship, such as a greater commitment, behavior oriented towards cooperation, increased trust,[15] and the perspective of a long-term relationship.[16]

Likewise, technological factors are considered to have a decisive role, insofar as they are regarded as an exogenous stimulus to enter cooperation agreements. Technology creates the appropriate environment for the development of cooperative behavior, due to the acceleration and speed of development of technological innovation, the rapid diffusion of these technologies, the shortening of product life cycles, and the intensive development of new associated products (Fernández & Arranz, 1999, p. 58).

In any case, we feel that it is important to understand cooperation as a process, which can be broken down into different stages according to a logical sequence: (1) the initial decision to cooperate, weighing up the possible benefits and risks; (2) the criteria for selecting the appropriate partner in the relationship; (3) the process of negotiation aimed at reconciling mutual expectations and establishing the most appropriate structure for the cooperation, according to its objectives, partners, risks, and so forth; (4) the management of the cooperation agreement as a new organization with its own objectives, and its own human and material resources to achieve them; (5) the evaluation of the tangible and intangible results of the cooperation, both for the firm and for its partners; and (6) the possible evolution of the agreement towards its maintenance, transformation, or dissolution (Menguzzato, 1992; Parkhe, 1993; Townsend, 2003).

The Role of IT in Internationalization Process

In the internationalization process, the function of IT has been explained according to three main lines of reasoning. First, on the basis of transaction cost economics (Williamson, 1975), IT can reduce the costs associated with international transactions, especially those generated as a consequence of managing the information required for the geographical expansion process, the design of safeguards in the international relationships, or for facilitating the process of negotiation that emerges in transnational transactions (Águila, Bruque, & Padilla, 2002). In this case the Internet has evolved into a revolutionary tool that is smoothing the path, and hence reducing the costs arising during the process of perfecting the transaction.

Second, IT, especially those technologies that act as support for IOS, has been used to identify or develop valuable and rare resources (Barney, 1991) necessary for the process of internationalization in firms. As has been pointed out in some previous studies (Medina & Ruiz, 2003), IT can be used to interconnect and activate many resources that are dispersed and not currently in use (dormant capabilities), which can become the basis for internationalization strategies. For example, internal networks, strategic information systems, knowledge maps, or enterprise resource planning (ERP) systems can identify resources (personnel specialized in internationalization processes, knowledge useful in international markets, information concerning the operation of international negotiation, etc.) that can prove to be key to the success of an internationalization strategy. This energizing capacity of IT is based on the ability to virtualize resources that are attributed to the latest generation of information technologies (Foster & Kesselman, 2004). The virtualization capacity involves eliminating the geographic variable that in traditional business impeded first the localization and then the sharing of geographically dispersed resources (manpower, data, information, or knowledge). Among the latest generation technologies enabling virtualization, we should highlight, are those that are based on the TCP/IP protocols (Internet, intranet, extranet, ERPs, etc.) as well as wide area networks that use more flexible modes of internal communication (the case of grid technologies).

Third, firms' possibilities of internationalization have been boosted precisely by the possibilities for internationalizing information systems (IS), thanks to the formation of international strategic networks (Gulati, Nohria, & Zaheer, 2000). In this case the internationalization processes on the one hand, and the creation of IOS on the other, can occur jointly and symbiotically. Thus, the creation of IOS on the supranational scale has its origin in a firm's decision to internationalize, or in the cooperation agreements between firms operating in different geographical areas. The development and expansion of these systems, combined with the emergence of strategic alliances, respond in turn to more far-reaching movements affecting economic and social structures around the world. Among these, we might mention the progressive opening up of national markets to foreign influence, the facilities for moving capital among countries, the standardization of the technologies permitting international interchange, and the convergence between cultures that were previously separate. In short, the establishment of international strategic networks together with the expansion of the IOS, which they entail, have been elements that have formed part of and had a role in the so-called globalization paradigm (Palvia, Palvia, & Whitworth, 2002).

Summarizing, the previous reflections allow us to say that IT has played an important role in corporate internationalization processes. However, much is still unknown about how the family nature of family firms can interfere in the internationalization processes that use some kind of strategic cooperation. This question, along with a description of the various modes of internationalization, will be examined in more detail in the following section.

The IT and the Internationalization of Family Firms through Strategic Alliances

Modes of Implementing an Internationalization Strategy

The literature on the choice of mode of entry in the host country[17] provides a wide range of types of penetration. Most authors agree that there are three general forms of implementing an internationalization strategy—exporting, licenses, and foreign direct investment (FDI)—considering the different intermediate levels that each includes as well as the possibility of embarking on the process individually or by means of cooperation agreements. There are authors, however, who also consider joint ventures as an alternative form of internationalization, studying these structures separately from the others (Tsang, 1997).

For the specific case of family firms and their idiosyncrasies, these firms may be strongly resistant to confiding certain intangible assets, upon which they base their competitive advantages, to foreign partners under the contractual form of licenses. We are referring to commercial secrets, know-how, brand, company name, and so forth. In this last case, for example, it could be regarded as being tantamount to ceding the family name itself. For all this, here we shall only consider the exporting and FDI modes of entry for this type of firm.

In Figure 2 we have summarized the various options for embarking on an internationalization process. Given the idiosyncrasies of family firms, the first question to ask when initiating an internationalization process is whether to do so individually or by means of some kind of cooperation relationship. Whatever the decision, exporting or direct investment has then to be considered. The objective of this current chapter is to analyze the use of strategic alliances in the internationalization processes of family firms as well as to determine the potential role IT may play in this process.

Figure 2. Modes of entry in family firm internationalization process

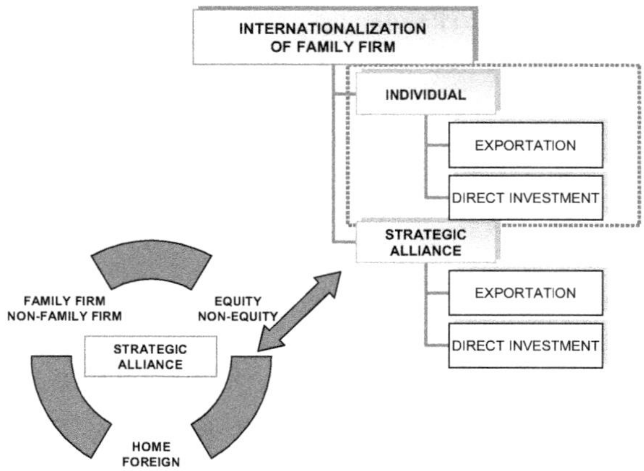

In this respect, we should consider the influence on the mode of entry chosen of whether the strategic alliance involves participation in the capital or not, whether the partners in the alliance are family firms or not, and whether the partners are from the home market or from the foreign market.

Firstly, family firms could reject cooperation agreements that involve participation in their capital, since this would mean ceding part of their control and even endangering the family nature and identity of the business. Of course, these firms could participate in the equity capital of the firms in the foreign country. Family firms tend to be characterized by their lack of financial resources (in many cases because they do not allow the entry of outside capital), and they could use the alliance to access the financial resources necessary to implement their internationalization strategy. Thus, the firms must weigh up the advantages of an equity alliance on the one hand, or seek alternatives to overcome the possible drawbacks this option entails. Faced with this dilemma, family firms are much more likely to prefer strategic alliances not involving participation in the capital.

Secondly, if the partner firm is a non-family firm each partner can benefit from the tangible and intangible resources that the other firm tends to possess. The non-family firm could obtain the intangible resources (know-how, experience, image, reputation, etc.) that constitute the strengths of family firms. These companies, in turn, could obtain the financial and technological resources that tend to be more common in non-family firms. Nevertheless, the majority of strategic alliances involving family firms tend to be with other family firms. An explanation for this fact could lie in the culture of this type of firm because there are values, commitments, and even shared strategies that encourage both the formation and the development of these alliances, which in most cases could be based on the relationships of trust between both firms.

Finally, we must bear in mind that the alliance can be formed in the firm's home market or in the target market. When implementing a strategy of internationalization, a fundamental resource is the firm's knowledge and experience of the market that it is targeting. Thus, a family firm that forms an alliance with a partner from the target market can exploit localization advantages (referred to in eclectic theory), which could speed up its international expansion. In this respect, it may be very beneficial to have family members living overseas, or to know other firms in different markets because of contacts maintained with them in previous generations. However, if the firm opts to form an alliance with another firm in the home market, with the aim of expanding both firms' international operations, the advantages gained will be different. In this case the risk and even the need for financing may be reduced, since the alliance partners can share certain resources (e.g., those that are necessary for the joint distribution of products in the international market).

The Influence of IT on Family Firm Internationalization via Strategic Alliances

As we saw in the previous section, family firms tend to choose two main forms of strategic alliance to undertake their internationalization processes. First, non-participative forms of cooperation, by means of which firms interested in international cooperation do not participate in the capital of any firm created *ad hoc*. Second, we can identify formulas requiring some level of participation in the capital of the firm set up to carry out the operations in the target country. In view of the particular characteristics of family firms—limited financial resources and conservative orientation in the running of the business—the formula chosen to enter new markets tends to be non-participative cooperation. But if there is no alternative to participative cooperation, family firms will seek a majority participation that ensures control of the relationship and preservation of the family nature of the business.

We can say that there are substantial differences in the objectives and scopes of the two formulas of cooperation proposed (participative and non-participative). However, both tend to consist of a number of stages that are required to carry out the cooperation and that have been mentioned earlier—initial decision, selection of partner, negotiation and structure, management, evaluation, and evolution. Although IT can have an overall generic impact on the cooperation process, we shall use an analytic criterion in our study of the functions that IT can realize, so we shall study the impact of IT on each stage of the cooperation process. In this discussion we shall also bear in mind the distinctive characteristics of participative and non-participative cooperation processes, since each option may have important implications for the way in which IT intervenes in family firms' internationalization processes.

Any process of international cooperation begins with the decision to cooperate, and those of family firms are no different. In this respect, IT can encourage firms in their desire to internationalize by forming an alliance. The various technological waves experienced at the end of the 20th century (data processing systems, generalization of personal computers, wide area networks, and the Internet) have had similar expansive effects to those experienced in the 19th and first half of the 20th century that were a consequence of the introduction of the railway, telegraph, and electricity (Peña, 2003). This effect can be summed up as an expansion of the physical, but also psychological, limits of firms' competitive frame of reference in countries where the technological change takes root. Technological intensification has also involved a standardization of the communication protocols between firms of different areas, aided by the progressive use of English as a *lingua franca* used in the technological interfaces (e-mail, web pages, forums, chats, etc.). The diffusion of networks based on TCP/IP (e.g., the Internet) has been an important catalyst for the internationalization of all types of firm, and this has been no less true for the family firm. At first, their conservative mentality and limited financial resources meant that family firms were slow to take up technological innovations (AECE, 2004). Subsequently, however, family firms have begun to avail themselves of the benefits provided by the new communication networks in terms of the greater transparency of target markets. This transparency, a consequence of the greater amount and quality of information provided through IT, can favor the internationalization of family firms. In addition, the increased transparency provided by IT has also encouraged family firms to consider the possibility of using cooperation to internationalize. In this process the growing number of vertical e-business models may have played a role. They have put firms from the same sector into contact, and hence helped them to identify potential partners.

Once it has been decided to use cooperation in order to internationalize, the firm needs to select a partner. We should point out, however, that in most cases partner selection occurs at the same time as the decision to cooperate itself. In spite of this, we have opted to differentiate the two stages in order to favor clarity in the analysis. The search for a partner in family firms' internationalization processes has been enormously facilitated by the appearance of Internet-based utilities, especially the world-wide-web (www) and e-mail, and the business models deriving directly from their application.[18] Specifically, intermediation models based on e-brokers and brokerage models, or vertical communities on the

Internet (Rappa, 2004), have increased the number of contacts between potential partners interested in internationalization strategies. With regards to family firms, as we have already noted, it is easier to find partners of a similar size and business culture. Nevertheless, the greater opportunities offered by this increased number of potential contacts cannot hide a number of drawbacks that are not new but that may be worsened by the possibilities opened up by IT. These disadvantages relate to the increasing probability of moral hazard or adverse selection—both a consequence of the opportunistic behavior that may be easier to hide in virtual environments than in the real world (Águila et al., 2002). Family firms may see this as an obstacle to their use of IT—at least their exclusive use—as it questions whether this is a reliable medium for finding an appropriate cooperation partner. Greater care will be needed when the cooperation formula chosen involves participation in the capital of the new firm, because of the increased risk that this entails. In the case of the family firm this strategy could even endanger the family nature of the firm.

The third stage of cooperation corresponds to the negotiation and structure. In this stage the partners must engage in a process of interchange aimed at aligning interests and arriving at an agreement that is satisfactory to both parties. Part of this agreement should detail the appropriate structure for the collaboration. IT can play an important role, both in the negotiation process and in the management of the government structure of the cooperation. In the negotiation, IT can provide the basis for the interchange of documents including the characteristics of the agreement. If the partners do not belong to the same country IT becomes essential during the negotiation (e-mail, videoconference, electronic data interchange, etc.). On the other hand, it can be even more complex to manage the interactions that may arise between the governance structure of the alliance and the configuration of the IS designed to link partners together and these to the new firm formed in the alliance. On this question, partners must come to an agreement about the ideal configuration of the IOS, as well as about the characteristics, levels of authorization, access levels, and so forth that will regulate the internal operation of the information system created as a consequence of the alliance. In this respect, choosing another family firm as partner could favor the process, in view of the similar organizational culture. An incorrect configuration of the new IOS can lead to deficiencies in the diffusion and access of information between the partners, a situation that in the worst case could damage the relationship of trust that any process of cooperation requires. In addition, family firms should pay particular attention to the steps that lead to the construction of the structure of the alliance, if they are not to be disadvantaged by the distribution in the power or access to the information generated by the *ad-hoc* information system created, especially when the cooperation involves both family and non-family firms.

IT will also play a key role in the day-to-day running of the alliance. Specifically, management is one of the areas where IT was first exploited in business. At present, IT, especially technologies relating to basic handling of documents, images, or business communications, is an essential element in any business relationship. The possibilities for improving management by using IT or communications technology have by no means been exhausted, and have been favored in the current stage of integration, in which previously separate computing tools are being combined. In this respect, ERP systems or customer relationship management (CRM) systems, together with the integration of networks in the form of extranets, intranets, or virtual private networks, are the basis upon which cooperation agreements can be technologically founded. Here, however, the family firm may find it more difficult to adapt because of its idiosyncrasies. Alternatively, family firms may therefore select a technologically more advanced partner, with greater experience in technology or a larger size, and this may help the cooperation relationship to exploit the management potential of IT more fully.

Finally, in the evaluation and evolution stage, again IT proves to be a useful tool. IT can help firms to obtain knowledge about the other partners and hence reduce the probability of opportunistic behavior. This capacity can be even more valuable for family firms, which generally place a high value on security in their cooperation initiatives, because of their higher risk aversion. This is even much more true if the operations are overseas. Under these circumstances an agency problem may arise (Jensen & Meckling, 1976), so that the partners in the cooperation are more likely to invest in control-oriented IT that reduces the risks of opportunistic behavior, fundamentally, when the partners are working with asymmetric information, the IS do not work in a consistent or coordinated way, or when a relationship of trust has not yet been consolidated. In these circumstances, the partners are more likely to invest in the acquisition and development of computer control systems such as executive information systems (EIS), balanced scorecards, or back-up systems with various levels of access.

Although, as we have mentioned earlier, IT can play an important role in each stage of the cooperation process, family firms may not be exploiting this tool to the full extent. The same elements that represent the advantages of family firms (commitment to the customer, consolidated traditional values) may, in some circumstances, have placed a brake on the use of IT as catalyst in the construction of strategic alliances. However, this situation, which may respond to a stereotypical view of family firm behavior, could be masking future trends and innovatory behaviors with respect to a greater openness to international cooperation and to the use of IT in the internationalization process. In Table 2 we provide a summary of the role of IT in each cooperation stage.

Table 2. The role of IT on strategic alliances among family firms

Stages in the cooperation process	Stage description	Role of IT	Unique features of family firms in each stage	Managerial implications to family firms for better implementing IT
Initial decision to cooperate	Choosing to act coordinated with and jointly versus other options thus achieving objectives that are supposed to be more accessible through collective action	The Internet may come out very useful to obtain information about the advantages/disadvantages of cooperation. The Internet also may be a correct option to know and imitate successful cooperation strategies run by other family firms.	Growth is a key issue in family firms in order to ensure the continuity of business. Cooperation is a worthy strategy to avoid the shortage of resources that family firms face out to.	The intensive use of IT may lead to cooperation among firms, making it easier to access information about potential partners and best practices in cooperation. Managers in family firms should directly promote systematic searching of best practices using IT-based tools.
Partner selection	Partner's profile is defined to better achieve cooperation goals	The Internet may be a suitable vehicle in looking for new partners. Among the electronic business models, vertical, virtual communities may constitute a worthy source of potential partners. Vertical virtual communities usually include search engines that may be useful for partner selection.	Family firms usually prefer other family firms with a similar culture as partners for cooperation. However, advantages and disadvantages of familiar character should be taken into account in relation to the specific resources being shared.	The tools provided by vertical, virtual communities may facilitate the search of new partners. These communities usually hold *project workshops* to help users for finding new partners. Managers in family firms should join these *project workshops* for finding suitable family firm partners.
Negotiation and structure	Potential partners interchange mutual expectations in order to achieve sought objectives, through cooperation. Partners also define the formal or informal cooperation structure, according to respective needs and expectations.	The new cooperation structure should be equipped with a parallel IT infrastructure. The interconnection among partners should be made in two levels. The first level deals with management of communication by means of standard communication devices such as e-mail, videoconferencing, forums, and so forth. The second level is the operative one. This deals with the elements in the value chain that should become transparent to other partners. Transparent elements in the value chain should be connected (at a transactional level or at scorecard level) with the partners' operational information systems.	Expectations and objectives to be achieved through cooperation may be different regarding the familiar vs. non-familiar character of the partner. Family firms usually develop more informal organizational structures than non-family firms. In these informal organizational structures, it is more common to have an overlapping between the roles played in the family and in the business. Furthermore, the informal organizational structure is usually related to less-developed planning and control systems. In this stage it may come out very useful to create a new organizational structure in order to better manage the agreement.	During the negotiation process partners should consider the costs related to the creation and maintenance of the new common information system, considering that these costs could be higher for family firms without suitable IS to interconnect. They also have to take into account the time needed to start the system. Other key issues that have to be addressed during negotiation are related to what information the partners are willing to share and the different access levels (open vs. limited access) for each type of partner. An inter-family task force or a steering committee may be useful to better manage this issue.

Table 2. continued

Stages in the cooperation process	Stage description	Role of IT	Unique features of family firms in each stage	Managerial implications to family firms for better implementing IT
Cooperation management	It is not enough to fix some objectives and design strategies to achieve them but the way of running them is a key point to achieving better results. Communication channels should be kept open along the whole cooperation's life and along all the organizational levels involved in the cooperation agreement.	Information systems created ad hoc to cooperate should be started. For that purpose, it may be useful to interconnect the transactional systems, as well as, databases involved in the cooperation agreement. The interconnection may be achieved based on a cooperative extranet that eases partners' access to shared systems. If cooperation is channelled through a new organization and if the partners are geographically distant, the creation of a liaison committee would be useful.	At this point of cooperation it is possible that family firms' managers have to struggle against reticences coming from some family members regarding the spreading of key information about the family firm among third parties. Lack of professionalization and the autocratic managerial style, which characterize some family firms, may be an obstacle to manage appropriately the cooperation relationship.	It is very likely that during the first steps of cooperation the system created, ad hoc, had performance inconsistencies or integrity problems. If the problems identified are relevant, it may appear a pessimistic feeling about the effectiveness of the IS. This negative, pessimistic phase usually disappears when initial adjustment problems among the partners' systems are sorted out. Family firms' features, such as flexibility, may drive them to face, with bigger success, adverse situations.
Evaluation	Expectations are compared with current data. The role of IS in this stage is relevant, particularly the information regarding the contributions (tangible and intangible) made by each partner to the common interest.	The cooperative extranet that is the basis for cooperation management may also be useful to create a joint balanced scorecard adapted to particular features of cooperation. Thus, firms could have a global view about each partner's contributions and results achieved through the cooperation relationship.	Cooperation objectives, in the case of family firms, can be evaluated not only regarding the short-term economic results but taking into account other variables related to a longer period, such as prestige, reputation, and so forth.	It is very important to create a consensus about the valuation of tangible and intangible contributions and results, because of the differences among the key resources of family and non-family firms. Evaluation of cooperation may become more objective and easier when family firms' management is run by professionals who do not belong to the family.
Evolution	As a consequence of the cooperation management and evaluation, the partners will decide about probable next steps regarding maintenance, transforming, or ending up the cooperation agreement	IT may improve the information used in the evolution stage. Besides, IT may also reduce the time needed to evaluate the relationship. Thanks to IT, firms involved in the cooperation agreement are in better condition to decide about the future of cooperation.	Along the cooperation process, the family firm usually realizes that cooperation has not ended up with loosing the familiar character of the firm; indeed, cooperation may have fostered business performance. Therefore, it is very likely that the family firm wants to keep the cooperation agreement in the future. The above idea is also backed by the family firms' ability to create strong long-term links. Family firms are more prone to create personal relationships with partners, so they may opt to create new relationships with other firms that take part of partners' personal and professional networks.	If cooperation is highly related to technology, managers should not be disappointed by negative results of IT implementation in the short-term. If the change related to IT has been relevant, it is likely not to obtain positive results for a period of six months to two years after the implementation. However, family firms will not evaluate the cooperation results in the short-term, so it is likely that cooperation evolves to fulfil each partner's expectations in the long-term.

Future Trends

With regards to the use of IT in the internationalization process of family firms, in the next few years we could experience various trends that may play a significant role in all the processes in which technology intervenes in some way or another. We are referring, on the one hand, to the process of technological integration, and on the other, to the globalization of technological change. Both of these processes are interrelated, and they have similar origins.

As far as the integration process is concerned, the current trends appear to indicate (Foster & Kesselman, 2004) that IT will increase its capacity of calculation, transmission, and storage of information exponentially, thanks to the sharing of disperse computer resources and telecommunications. These new technologies, which will mark a new stage in the Internet, will be organized around the so-called grid technologies. Thus, in the near future—conceivably in the next five years—firms will be able to access computing resources (software, storage capacity, data processing, and calculation) ubiquitously, using standard communication protocols that are globalized at the international level. These protocols will, moreover, have the advantage of being conceived using the *open-source* system, so that firms, in general, and family firms, in particular, will be able to use these new systems more easily.

It is clear that global, collaborative, and open-source tools may be ideal techniques to enable family firms with limited resources (particularly financial and technological resources) to embark on internationalization with more chance of success. In particular, family firms can exploit the advantages of the new technological changes, since the relative disadvantages brought about by the complexity of use of the technology may well be mitigated in the future. This trend could be accentuated if at the same time the family firm embarks on internal change, such as accepting outside capital and changing attitudes unreceptive to the formation of strategic alliances.

In this respect, IT can contribute to the creation of a virtual network and platform of cooperation between family firms, aimed at encouraging the internationalization of firms less able to evaluate or contemplate this process individually. The opportunity to access information about previous experiences, markets, and partners could help reduce the risk of this strategy failing.

By making use of the possibilities offered by IT, it will be easier to elaborate censuses of family firms and categorize them into various types (in function of the proportion of the family's share of the equity capital, of family members on the board of directors, or of family members occupying management posts) at the national and international level. These will serve as directories that firms can refer to when seeking potential partners for cooperation agreements, helping to ensure a close cultural match between the parties.

Conclusions

The theories of internationalization under the economic perspective or the resource-based view explain firm internationalization, among other motivations, on the basis of the possession of certain ownership advantages relating to intangible resources. This type of resource constitutes one of the strengths of family firms and hence could be considered as sufficient motive to embark on international expansion. But in addition family firms could find that the desire to expand the business or the need to diversify risk also motivates carrying out geographical diversification of their markets. However, before implementing this strategy family firms would be advised to address some of the weaknesses that characterize them, such as, for example, their lack of technological and financial resources.

On the other hand, apart from the influence that the strengths and weaknesses of this type of firm can play on their internationalization processes, there are other aspects that can also favor or limit this strategy. Thus, a desire to diversify risk and an aim to expand the business can encourage this process of internationalization. In contrast, the lack of knowledge and poor preparation of the family (especially in international business); the management style; the management's reluctance to form alliances; and the existence of relatively undeveloped information and control systems can all be considered obstacles to the international strategy.

The mode of entry chosen can also be affected by the particular nature of this type of firm. Thus, when the firm opts to cooperate with other firms to expand its market internationally, the family nature or otherwise of the partner firm, whether the alliance involves participation in the capital or not, and whether the alliance is undertaken in the home country or in the target country overseas, are all relevant aspects for these firms' internationalization strategies.

Analysis of cooperation as the means towards internationalization is enriched when we consider it from the perspective of a process, going through various phases, in each of which attention is centered on different aspects that can influence the success or failure of the cooperation experience. This process perspective enables us more deeply to understand the role potentially played by IT as a facilitating element of the cooperation, in this case at the service of family firms' internationalization, allowing both cooperation and internationalization to be carried out transparently, safely, and reliably.

With regards to IT's impact in the different stages of the cooperation process, we can say that these technologies have had a growing influence in practically all the stages. Technological development and the appearance of international networks have encouraged a growing number of contacts with potential partners. There has also been a proliferation in the information available about each partner, improving the chances of successful selection. The biggest doubts arise perhaps in the negotiation and structure stage, since problems of harmony may emerge between the cooperation structures and the structures of the IS upon which the cooperation is founded. Finally, IT has had, and indeed continues to have, a substantial effect in the management, evaluation, and evolution stages.

Whether family firms can successfully exploit the benefits of IT to the full extent, will depend on whether they can carry out a series of internal adjustments. These firms need to become more open to national and international cooperation, obtain more financial resources, and connect with technologically strategic networks in the various geographical areas. Family firms should look for increasing their technological resource allocation through individual or combined investments; taking part in partnerships and associations; and encouraging technological relationship with suppliers and clients in order to facilitate the technological *contagion*. In this way, family firms can adapt themselves more closely to the changes that are likely to occur in the future as a result of the technological integration processes currently underway.

Family's members should show an open attitude to develop an internationalization strategy based on the use of IT. Therefore, it should be advisable that the family's members, who join up with the board of directors or management team of the company, have been trained abroad, have learned by the previous experience of having worked in multinational enterprises, have often travelled abroad due to commercial aims, and are comfortable with implementing IT into their training and the development of their professional activity as well.

Family firms should employ IT to extend their business view, improving their knowledge of the host markets in order to analyze their potential, their possibilities of succeeding, and their potential partners to ally with. Likewise, it would be advisable to get information about other similar internationalization experiences to try to find the key aspects in the process. In this respect, firms should plan when and how to internationalize their business activities—fixing objectives, business scheme—according to the firms' resources and capabilities and their general development strategy.

The employment of IT will allow family firms to dispose of information about previous experiences of managerial cooperation, together with the opportunity to spread their brand name and to extend their standing on an international-wide basis, in order to encourage their entry into host markets playing the role of the potential partner with which other firms are willing to collaborate with. Thus, family firms should give more information about themselves, for instance, through their Web site, where you should find the firm's history; members; philosophy and objectives; products and/or services' catalogue; distributors in the domestic and international markets; news and press releases; bonds with other public and/or private institutions; and so forth, which encourages the establishment of cooperation agreements to expand commercial activity to different markets.

The family firm has to generate, about the internationalization project through strategic alliances, a perception that the firm has bet solidly on internationalization and cooperation as the best way to get it, not only as a second-best to convince potential partners in order to encourage their commitment.

It should be established clearly, the resources and capabilities are disposed by the firm, pointing out those which are potentially more useful and deciding if the firm is prepared to bring and share them within the cooperation relationship. In this respect, it also should be laid out how and when the expected contributions made by the rest of the partners will take place.

Related to the management of the cooperation relationships, established to become internationalized, family firms have to take into account the need for modifying their managerial style towards a pattern based on negotiation and on the searching of consensus, instead of one based on ownership and decisions based on authority, without losing the familiar character of the firm. The incorporation of non-family professional members into the family firms' boards of directors and/or into the managerial team could bring objectivity and technological knowledge, which can be very useful to improve the possibilities of succeeding in the internationalization project through strategic alliances.

The IT offer an open fan of possibilities to provide the regular interconnection among the involved parts in the cooperation agreement, up to the point that information can pass among the partners in real time through the common informational system of the partnership. The executive management of the partnership could be set up through a virtual steering committee, which allows making joint decisions. This is especially important when those subjects where partners's participation improve their knowledge about the current situation and their motivation and commitment to collaborate.

To sum up, family firms should take advantage of the opportunities offered by public institutions, business associations, trade chambers, and so forth, to promote the use of new technologies, by means of training and financial policies. Besides, there are different public programs to support internationalization processes and to promote cooperation agreements. The presence, preference financing, subsidies, and advice by the technicians of these institutions, can become a useful tool to support the beginning of IT, internationalization, and cooperation implementation processes. So, family firms should encourage their relationship with other firms and/or public institutions (taking part in international trade fairs, public actions, being an active agent in different events, etc.).

This theoretical analysis of the influence of IT in family firms' internationalization processes through cooperation agreements is necessarily of a speculative nature, opening up possibilities of interpretation and generating more questions than answers. However, we understand that this fact, far from being a weak point of this work—although undoubtedly a limitation—is in fact, a strength in that it stimulates thought and scientific debate, in an attempt to find new answers that contribute to a greater understanding of the reality and competitiveness of family firms that decide to internationalize by means of cooperation, in an environment characterized by a continuing technological revolution.

References

AECE. (2004). *Estudio sobre comercio electrónico B2C 2004.* Madrid: Asociación Española de Comercio Electrónico. (*Study on electronic commerce B2C 2004.* Madrid: Spanish Association of Electronic Commerce.)

Águila, A. R., Bruque, S., & Padilla, A. (2002). Global information technology management and organizational analysis: Research issues. *Journal of Global Information Technology Management, 5*(4), 18-37.

Amat, J. (1998). *La continuidad de la empresa familiar.* (*The continuity of family firms.*) Barcelona: Gestión 2000.

Andersen, O., & Kheam, L. S. (1998). Resource-based theory and international growth strategies: An exploratory study. *International Business Review, 7*(2), 163-184.

Anderson, E., & Gatignon, H. (1986). Modes of foreign entry: A transaction costs analysis and propositions. *Journal of International Business Studies, 17*(3), 1-26.

Ariño, A., Cappuyns, K., Gallo, M., & Máñez, I. (2000, September 21-22). El proceso de internacionalización avanzado de las empresas familiares: Involucración de las empresas familiares en alianzas estratégicas. (The advanced internationalization process of family firms: Family firms involvement in strategic alliances.) In *Proceedings of I Congreso Nacional de Investigación sobre la Empresa Familiar, Instituto de Investigación de la Empresa Familiar* (pp. 375-399). Valencia, Spain: Organismo Público Valenciano de Investigación (OPVI).

Barber, B. (1983). *The logic and limits of trust.* New Brunswick, NJ: Rutgers University Press.

Barnes, L. B., & Hershon, S. A. (1976). Transferring power in the family business. *Harvard Business Review, 53*(4), 105-114.

Barney, J. B. (1991). Firm resources and sustained competitive advantage. *Journal of Management, 17*(1), 99-120.

Beamish, P. W. (1988). *Multinational joint ventures in developing countries.* London-New York: Routledge.

Beckhard, R., & Dyer, W. G. (1983). Managing continuity in the family-owned business. *Organizational Dynamics, 12*(1), 5-12.

Buckley, J., & Casson, M. C. (1976). *The future of the multinational enterprise.* London: McMillan.

Burack, E. H., & Calero, C. M. (1981). Seven perils of family business. *Nation's Business, 69*(1), 62-64.

Calder, G. H. (1961). The peculiar problems of a family business. *Business Horizons, 4,* 93-102.

Cauffman, L. (1996). El desafío de la empresa familiar: Sus relaciones. (The family firm challenge: Networking.) In M. A. Gallo (Ed.), *La empresa familiar 5* (*The family firm 5*) (pp. 99-113). *Publicaciones de la Cátedra de Empresa Familiar.* Barcelona, Spain: University of Navarra, IESE.

Caves, R. E. (1971). International corporations: The industrial economics of foreign investment. *Economica, 38,* 1-27.

Claver, E., Molina, J., & Quer, D. (2000, September 21-22). Un enfoque estratégico para la internacionalización de la empresa familiar. (A strategic rationale for the internationalization in family firms.) In *Proceedings of I Congreso Nacional de Investigación sobre la Empresa Familiar, Instituto de Investigación de la Empresa Familiar* (pp. 419-447). Valencia, Spain: Organismo Público Valenciano de Investigación (OPVI).

Coleman, S., & Carsky, M. (1999). Sources of capital for small family-owned businesses: Evidence from the National Survey of Small Business Finances. *Family Business Review, 12*(1), 73-85.

Collis, D. J. (1991). A resource-based analysis of global competition: The case of the bearings industry. *Strategic Management Journal, 12,* 49-68.

Contractor, F. J., & Lorange, P. (Eds.). (1988). *Cooperative strategies in international business.* New York: Lexington Books.

Danco, L. (1980). *Inside the family business.* Cleveland: The University Press.

Davis, P. (1983). Realizing the potential of the family business. *Organizational Dynamics, 12*(1), 47-56.

Davis, P., & Harveston, P. (2000). Internationalization and organizational growth: The impact of internet usage and technology involvement among entrepreneur-led family business. *Family Business Review, 13*(2), 107-120.

DGPYME. (2005). Guía básica de cooperación entre empresas, *el mejor camino para la internacionalización de las PYME*. Retrieved March 7, 2006, from http://wsie.ipyme.org/IPYME/es-ES/CooperacionEmpresarial/Publicaciones

Donckels, R., & Fröhlich, E. (1991). Are family business really different? European experiences from STRATOS. *Family Business Review, 4*(2), 149-160.

Donckels, R., & Lambrecht, J. (1999). Are family businesses really different? What we know from western European business research that could be applied to the reemergence of family-based enterprises in east central Europe. *Family Business Review, 12*(2), 171-191.

Donnelley, R. G. (1964). The family business. *Harvard Business Review, 42*(4), 93-105.

Doz, Y. L., & Hamel, G. (1998). *Alliance advantage. The art of creating value through partnering.* Boston, MA: Harvard Business School Press.

Dunning, J. H. (1979). Explaining changing patterns of international production: In defense of the eclectic theory. *Oxford Bulletin of Economics and Statistics, 41,* 269-295.

Dunning, J. H. (1981). *International production and the multinational enterprise.* London: Allen & Urwin.

Dunning, J. H. (1988). The eclectic paradigm of international production: A restatement and some possible extensions. *Journal of International Business Studies, 19,* 1-31.

Dunning, J. H. (1992). *Multinational enterprises and the global economy.* Wokingham, UK: Addison-Wesley.

Dunning, J. H. (1997). *Alliance capitalism and global business.* London-New York: Routledge.

Durán Herrera, J. (2001). *Estrategia y economía de la empresa multinacional.* (*Strategy and economics of family firms.*) Madrid: Pirámide.

Dyer, J. H., Kale, P., & Singh, H. (2001). How to make strategic alliances work. *MIT Sloan Management Review, 42*(4), 37-43.

Dyer, W. G. (1986). *Cultural change in family firms: Anticipating and managing business and family traditions.* San Francisco: Jossey-Bass.

Faulkner, D. O., & de Rond, M. (2000). Perspectives on Cooperative Strategy. In D .O. Faulkner & M. de Rond (Eds.), *Cooperative strategy: Economic, business and organizational issues* (pp. 3-39). Oxford, UK: Oxford University Press.

Fernández, E. (1996). *Innovación, tecnología y alianzas estratégicas. Factores clave de competitividad.* (*Innovation, technology and strategic alliances: Key factors for success.*) Madrid: Civitas.

Fernández, J. C., & Arranz, N. (1999). *La cooperación entre empresas: Análisis y diseño.* (*Co-operation among firms: Analysis and design.*) Madrid: ESIC.

Foster, I., & Kesselman, C. (2004). *The Grid2: Blueprint for a new computing infrastructure.* San Francisco: Morgan Kaufman-Elsevier.

Fuentes, G., Martínez, R., & Vallejo, M. C. (2004, September 8-10). La empresa familiar: Una revisión de la literatura. (Family firms: A literature review.) In *Proceedings of XIII International Conference AEDEM, Cartagena de Indias* (pp. 217-244). Cartagena de Indias, Colombia: The European Association of Management and Business Economics.

Fuentes, G., Vallejo, M. C., & Martínez, R. (2005, February 2-4). El proceso de internacionalización de la empresa familiar. (The internationalization process in family firms.) In E. P. Díez de Castro & M. Rey Moreno (Eds.), *Proceedings of XV Spanish-Portuguese Meeting of Scientific Management* (pp. 167-183). Sevilla, Spain: University of Seville. Mergablum.

Gallo, M. (1995). *Empresa familiar: Textos y casos.* (*Family firms: Text and cases.*) Barcelona: Editorial Praxis.

Gallo, M. (2002, June 28-35). *Unidad y confianza en la empresa familiar* (pp. 28-35). University of Navarra, IESE. Retrieved March 7, 2006, from http://www.ee-iese.com/86/pdf86/af3.pdf

Gallo, M., & Cappuyns, K. (1999). La internacionalización de la empresa familiar. El papel de las alianzas estratégicas (The internationalization of family firms. The role of strategic alliances.) *Iniciativa Emprendedora y Empresa Familiar, 16,* 38-44.

Gallo, M., & García, C. (1989). La empresa familiar en la economía Española. (The family firm in the Spanish economy.) *Papeles de Economía Española, 39,* 67-85.

Gallo, M., & García, C. (1996). Important factors in family business internationalization. *Family Business Review, 9*(1), 45-60.

Gallo, M., & García, C. (1998). La proyección internacional de la empresa familiar. Factores impulsores y retardadores. *Iniciativa Emprendedora y Empresa Familiar, 11,* 41-45. (The international projection of family firms. Fostering and hindering factors. *Entrepreneurship and Family Firms, 11*, 41-45.)

Gallo, M. & Luostarinem, R. (1993). Internacionalización: Un Cambio Desafiante para las Empresas Familiares. (Internationalization: A challenging change for family firms.) In M. A. Gallo (Ed.), *La empresa familiar 4. Estudio y ediciones* (*The family firm 4. Study and editions*) (pp. 91-101). Barcelona, Spain: University of Navarra, IESE.

Gallo, M., & Sveen, J. (1991). Internationalizing the family business: Facilitating and retraining factors. *Family Business Review, 4*(2), 181-190.

Gannon, M. (1993). Towards a composite theory of foreign market entry mode choice: The role of marketing strategy variables. *Journal of Strategic Marketing, 1*(1), 41-54.

Gilbert, N. (1989). Can a family business survive mergermania and divorce? *Management Review, 78*(1), 38-42.

Ginebra, J. (1997). *Las empresas familiares.* (*Family firms.*) Mexico: Panorama.

Glaister, K. W., & Buckley, P. J. (1996). Strategic motives for international alliance formation. *Journal of Management Studies, 33*(3), 301-332.

Glaister, K. W., Husan, R., & Buckley, P. J. (2004). *Strategic business alliances. An examination of the core dimensions.* Cheltenham, UK: Edward Elgar.

Gomes-Casseres, B. (1996). *The alliance revolution.* Cambridge, MA: Harvard University Press.

Grant, R. (1991). The resource-based theory of competitive advantage: Implications for strategy formulation. *California Management Review, 33*(3), 114-135.

Guillén, M. (2000). Business groups in emerging economies: A resource-based view. *Academy of Management Journal, 43*(3), 362-380.

Guillén, M. (2001). International business (pp. 7768-7771). In *International Encyclopedia of the Social and Behavioral Sciences.* Oxford: Elsevier Publishers.

Gulati, R., Nohria, N., & Zaheer, A. (2000). Strategic networks [Special Issue]. *Strategic Management Journal, 21*(3), 203-215.

Gupta, A. K., & Govindarajan, V. (1991). Knowledge flows and the structure of control within multinational corporations. *Academy of Management Review, 16*(4), 768-792.

Hall, R. (1992). The strategic analysis of intangible resources. *Strategic Management Journal, 13*(2), 135-144.

Hamel, G. (1991). Competition for competence and inter-partner learning within international strategic alliances. *Strategic Management Journal, 12,* 83-103.

Handler, W. C. (1989). *Managing the family firm succession process: The next-generation family member's experience.* UMI Dissertation Services.

Harris, D., Martínez, J. L., & Ward, J. L. (1994). Is strategy different for family-owned businesses? *Family Business Review, 7*(2), 159-176.

Hayes, J. L. (1981). All in the family. *Management Review, 70*(7), 4-7.

Heck, R. K. Z., & Scannell, E. (1999). The prevalence of family business from a household sample. *Family Business Review, 12*(3), 209-224.

Hennart, J. F. (1982). *A theory of the multinational enterprises.* Ann Arbor, MI: The University of Michigan Press.

Hill, C. W. L., Hwang, P., & Kim, C. W. (1990). An eclectic theory of the choice of international entry mode. *Strategic Management Journal, 11*(2), 117-128.

Holtbrügge, D. (2004). Management of international strategic business cooperation: Situational conditions, performance criteria and success factors. *Thunderbird International Business Review, 46*(3), 255-274.

Horton, T. R. (1986). Managing in a family way. *Management Review, 75*(2), 3.

Hymer, S. (1976). *The international operations of national firms: A study of direct foreign investment.* Cambridge, MA: The MIT Press.

Ibrahim, A., & Ellis, W. (1994). *Family business management concepts and practice.* Dubuque: Kendall/Hunt Publishing Company.

James, H. S. (1999). What can the family contribute to business? Examining contractual relationships. *Family Business Review, 12*(1), 61-72.

Jensen, M. C., & Meckling, W. H. (1976). Theory of the firm: Managerial behaviour, agency costs and ownership structure. *Journal of Management, 25*(3), 305-360.

Johanson, J., & Vahlne, J. E. (1977). The internationalization process of the firms-A model of knowledge development and increasing foreign market commitment. *Journal of International Business Studies, 8*(1), 23-32.

Johanson, J., & Vahlne, J. E. (1990). The mechanism of internationalization. *International Marketing Review, 7*(4), 11-24.

Kaslow, F., & Kaslow, S. (1993). The family that works together: Special problems of family business. In S. Zedeck (Ed.), *Work, families and organizations* (pp. 312-351). San Francisco: Jossey-Bass.

Kepner, E. (1983). The family and the firm: A coevolutionary perspective. *Organizational Dynamics, 12*(1), 57-70.

Kets de Vries, M. (1993). Lo bueno y lo malo de las empresas de titularidad familiar. (Advantages and disadvantages of family-owned firms.) *Harvard Deusto Business Review, 4,* 32-44.

Kets de Vries, M. (1996). *Family business: Human dilemmas in the family firm.* London: Thomson Business Press.

Killing, J. P. (1983). *Strategies for joint venture success.* New York: Praeger.

Klein, S. B. (2000). Family businesses in Germany: Significance and structure. *Family Business Review, 13*(3), 157-181.

Leach, P. (1993). *La empresa familiar.* (*Family firms.*) Barcelona: Granica.

Levinson, H. (1971). Conflicts that plague family businesses. *Harvard Business Review, 49*(2), 90-98.

Luostarinen, R., & Hellman, H. (1994). *The internationalization processes and strategies of Finnish family firms.* CIBR Research Paper Series Y-1. Helsinki: Helsinki School of Economics.

Madhok, A. (1997). Cost, value and foreign market entry mode: The transaction and the firm. *Strategic Management Journal, 18*(1), 39-61.

Márquez-García, A. M., & Casani, F. (2001, April 19-21). Trust: The newest trend in co-operation relationships study. In J. E. Ricart (Ed.), *Proceedings of Founding Conference of the European Academy of Management (EURAM). European Management Research. Trends and Challenges.* Barcelona, Spain: European Academy of Management & IESE Business School.

Medina, J. A., & Ruiz, J. (2003). IT and dynamic capabilities. *International Journal of Information Technology, 9*(1), 47-66.

Menguzzato, M. (1992). *La cooperación empresarial. Análisis de su Proceso.* Valencia: IMPIVA. Instituto de la Mediana y Pequeña Industria Valenciana: Generalitat Valenciana. (*Co-operation among firms: A process analysis.* Valencia: IMPIVA. Institute of Valencian Small and Medium Firms.)

Miller, E. J., & Rice, A. K. (1988). The family business in contemporary society. *Family Business Review, 1*(2), 193-210.

Morck, R., & Yeung, B. (2003). Agency problems in large family business groups. *Entrepreneurship: Theory and Practice, 27*(4), 367-382.

Ohmae, K. (1989). The global logic of strategic alliances. *Harvard Business Review, 67*(2), 143-155.

Okoroafo, S. (1999). Internationalization of family business: Evidence from northwest Ohio, U.S.A. *Family Business Review, 12*(2), 147-158.

Palvia, P. C., Palvia, S. C. J., & Whitworth, J. E. (2002). Global information technology: A meta analysis of key issues. *Information and Management, 39*(5), 403-414.

Parkhe, A. (1993). Strategic alliance structuring: A game theoretic and transaction cost examination of interfirm cooperation. *Academy of Management Journal, 36*(4), 794-829.

Peña, J. (2003). *Historias de las telecomunicaciones.* (*Stories on telecommunications.*) Barcelona: Ariel.

Peng, M. W. (2001). The resource-based view and internacional business. *Journal of Management, 27*(6), 803-829.

Perlmutter, H. V., & Heenan, D. A. (1986). Cooperate to compete globally. *Harvard Business Review, 64*(2), 136-152.

Pla, J. (2000). *La estrategia internacional de la empresa Española. (The international strategy of Spanish firms.)* Vall d'Albaida, Ontinyent: Fundació Universitària.

Porter, M. E. (1990). *The competitive advantage of nations.* New York: Macmillan.

Poutziouris, P. Z. (2001). The views of family companies on venture capital: Empirical evidence from the UK small to medium-size enterprising economy. *Family Business Review, 14*(3), 277-291.

Poza, E. (1995). *A la sombra del roble: La empresa privada familiar y su continuidad. (Under the oak's shadow: Private family firms and their continuity.)* Ohio: Editorial Universitaria para la Empresa Familiar.

Rappa, M. A. (2004). The utility business model and the future of computing services. *IBM Systems Journal, 43*(1), 32-43.

Root, F. (1987). *Foreign market entry strategies.* New York: AMACOM.

Root, F. (1994). *Entry strategies for international markets.* New York: Lexington Books.

Rubio, E. A. (2004). *La internacionalización de la Pyme Europea: Génesis y evolución de las políticas de apoyo a la cooperación empresarial en el marco de la Unión Europea. (The internationalization of the European SMEs: Genesis and evolution of the policies for co-operation promotion in the European Union.)* Granada: Universidad de Granada.

Salganicoff, M. (1990). Women in family businesses: Challenges and opportunities. *Family Business Review, 3*(2), 125-137.

Sarkar, M., & Cavusgil, S. T. (1996). Trends in international business thought and literature: A review of international market entry mode research: Integration and synthesis. *The International Executive, 38*(6), 825-847.

Seijo, E., & Menéndez, S. (2004, September 19-21). Internacionalización de las grandes empresas familiares españolas. (Internationalization in big Spanish family firms.) In *Proceedings of XIV ACEDE Conference, Conocimiento y Competitividad.* Murcia, Spain: Scientific Association of Business Economy and Management (ACEDE).

Shanker, M. C., & Astrachan, J. H. (1996). Myths and realities: Family businesses' contribution to the U.S. economy: A framework for assessing family business statistics. *Family Business Review, 9*(2), 107-123.

Sharma, P. (2004). An overview of the field of family business studies: Current status and directions for the future. *Family Business Review, 17*(1), 1-36.

Stern, M. (1986). *Inside the family-held business.* New York: Harcourt Brace Jovanovich.

Tagiuri, R., & Davis, J. A. (1996). Bivalent attributes of the family firm. *Family Business Review, 9*(2), 199-208.

Tallman, S. (1991). Strategic management model and resource-based strategies among MNEs in a host market. *Strategic Management Journal, 12,* 69-82.

Tallman, S. (1992). A strategic management perspective on host country structure of multinational enterprises. *Journal of Management, 18*(3), 455-471.

Tallman, S., & Fladmoe-Lindquist, K. (2002). Internationalization, globalization, and capability-based strategy. *California Management Review, 45*(1), 116-135.

Townsend, J. D. (2003). Understanding alliances: A review of international aspects in strategic marketing. *Marketing Intelligence & Planning, 21*(3), 143-155.

Tsang, E. (1997). Choice of international technology transfer mode. *Management International Review, 37*(2), 151-168.

Tsang, E. (2000). Transaction cost and resource-based explanations of joint ventures: A comparison and synthesis. *Organization Studies, 21*(1), 215-242.

Tsang, E. (2002). Learning from overseas venturing experience. The case of Chinese family businesses. *Journal of Business Venturing, 17*(1), 21-40.

Vallejo, M. (2003). *La cultura de la empresa familiar como ventaja competitiva: Una aplicación al sector de concesionarios de automoción.* Tesis doctoral no publicada. Departamento de Administración de Empresas, Contabilidad y Sociología, Universidad de Jaén. (*The culture of family firms as a source of competitive advantage.* Unpublished doctoral thesis. Department of Business Administration, Accounting and Sociology, University of Jaén.)

Ward, J. L. (1987). *Keeping the family business healthy: How to plan for continuing growth, profitability, and family leadership.* San Francisco: Jossey-Bass.

Williamson, O. E. (1975). *Markets and hierarchies: Analysis and antitrust implications. A study in the economics of internal organization.* New York: Free Press.

Endnotes

1 In the case of the U.S., for some authors (Beckhard & Dyer, 1983; Dyer, 1986; Ibrahim & Ellis, 1994; Kaslow & Kaslow, 1993; Stern, 1986; Ward, 1987) these firms represent 90% of all companies, generate 50% of GNP (Coleman & Carsky, 1999; Salganicoff, 1990; Shanker & Astrachan, 1996) and employ 50% of the active population (Barnes & Hershon, 1976; Danco, 1980; Ibrahim & Ellis, 1994; Ward, 1987). In the European Union, family firms are a majority in almost all member countries (Donckels & Fröhlich, 1991). For the case of Spain, these firms represent between 65% and 80% of the business structure, employ 65% of the active population and generate between 50% and 65% of the GDP (Amat, 1998; Gallo & García, 1989).

2 We are referring to the recent growth of various institutions (Family Business Network, the Spanish Family Firm Institute, associations of family firms and university chairs), which study the particular issues involved around this type of firm in an attempt to improve and develop them.

3 We might mention the work of Sharma (2004) and Fuentes, Martínez, & Vallejo (2004), mainly after reviewing the following journals: Family Business Review, Entrepreneurship: Theory and Practice, Journal of Business Venturing, Journal of Small Business Management, and Harvard Business Review.

4 Kets de Vries (1993, 1996); Ibrahim and Ellis (1994); Ginebra (1997); and Donckels and Lambrecht (1999).

5 This classification is based on a review of the work of Calder (1961); Donnelley (1964); Levinson (1971); Burack and Calero (1981); Hayes (1981); Davis (1983); Kepner (1983); Horton (1986); Miller and Rice (1988); Gilbert (1989); Kets de Vries (1993, 1996); Leach (1993); Ibrahim and Ellis (1994); Gallo (1995, 2002); Poza (1995); Cauffman (1996); Tagiuri and Davis (1996); Ginebra (1997); Donckels and Lambrecht (1999); S. James (1999); and Poutziouris (2001).

6 Some of the work studying these aspects includes Gallo and Sveen (1991); Gallo and Luostarinem (1993); Luostarinen and Hellman (1994); Gallo and García (1996, 1998); Gallo and Cappuyns (1999); Okoroafo (1999); Ariño, Cappuyns, Gallo, and Máñez (2000); Claver, Molina, & Quer (2000); Davis and Harveston (2000); Tsang (2002); and Seijo and Menéndez (2004).

7 Gupta and Govindarajan (1991); Tallman (1992); Madhok (1997); Peng (2001); and Tallman and Fladmoe-Lindquist (2002).

8 Tallman (1991); Collis (1991); Hamel (1991); Andersen and Kheam (1998); and Guillén (2000).

9 So, family firms are more inclined to develop trust-based relationships, not only with suppliers and clients, but also with other partners in strategic alliances, especially when family firms have kept long-term relationships with potential partners. Trust, as a special value in family firms, may have, otherwise, negative consequences for strategic alliances, since family firms can reject potential partners that had not had a previous trustworthy relationship with them. The need to set up a strong, long-term link with partners may, therefore, prevent family businesses from arising new strategic alliances with relatively non well-known partners, even though when the partners may, eventually, be suitable for cooperation. Thus, these firms are more proud of their trade name and brands, both obtained thanks to former generations, than non-family firms. So, it could be one of the motives to be willing to expand their activities in markets different to the domestic ones. However, this pride could have a negative influence on the way of developing this strategy through strategic alliances, since this could imply the sharing of some resources that could even be its trade name, brand, prestige, image, and standing with other firms which have not contributed to family firms' effort to get all above.

10 "Alliances appear to be most common among second-tier competitors or firms trying to catch up ... (while) global leaders rarely if ever rely on a partner for assets and skills essential to competitive advantage" (Porter, 1990, p. 67). Nowadays, according to a study by Dyer, Kale, and Singh (2001), the 500 largest firms in the world have an average of 60 major cooperation agreements each.

11 In Spain, the Directorate General for SME Policy recommends that when firms, particularly small and medium-sized enterprises (SMEs), decide to internationalize or enter new markets, they do so in company, seeking cooperation with other national or international firms that share their business philosophy (DGPYME, 2005).

12 According to Perlmutter & Heenan (1986), only firms that cooperate across national borders will be able to meet these new challenges and compete globally.

13 The role of learning through cooperation can be a key element in many international relationships (Gomes-Casseres, 1996).

14 Empirical studies demonstrate that dominant parent structures have lower failure rates than shared management structures (Beamish, 1988; Killing, 1983).

15 As Barber (1983, p. 21) points out, trust allows the parties to express and maintain the shared values upon which the organization is founded. Trust acts as an integrating mechanism that creates and sustains solidarity in

relationships and social systems. Thus, trust has become a central issue for understanding cooperation relationships (Márquez-García & Casani, 2001).

16 Many theories may explain why international strategic partnerships are concluded, but they do not provide much evidence of how to manage them (Holtbrügge, 2004, p. 261).

17 Root (1987, 1994); Anderson and Gatignon (1986); Sarkar and Cavusgil (1996); Pla (2000); and Durán (2001).

18 These utilities allow firms to keep in contact and build the trust that characterizes this type of firm.

Chapter IX

AVEA: A Merger of Two Large GSM Operators in the Turkish Telecommunications Sector

Yonca Gürol, Yildiz Technical University, Turkey

Özen Aşık, Yildiz Technical University, Turkey

Abstract

During the past 10 years, emerging factors such as globalization, rapid technological developments, and limited resources have created a new competitive landscape for firms. These developments forced companies to adopt several strategies to survive and excel. One of these adaptive strategies was to form alliances between companies to manage both threats and opportunities in this changed business environment. This study focuses on the merger between two GSM operators in Turkey. The newly established firm, AVEA, is an example of a successful merger in the Turkish telecommunications sector that was well thought-out, efficiently designed, and realized within a year. The most important ingredient in this process was the change in information technology (IT) infrastructures. This study, primarily concerned with the structural changes in the IT department, presents a detailed analysis of the merger process's different aspects and comments based on literature review about information and communication technologies.

Introduction

The business world today has been witnessing a greater pace and more intense competition, due to various environmental changes that force organizations to react more rapidly and flexibly. The most pervasive of these changes include globalized markets, larger investment needs, and accelerated technological developments. Developments in the technological arena especially put organizations under pressure to work on matching strategies that help them to sustain their long-run existence in an environment of technological convergence.

Strategic management is concerned with the effective and efficient use of organizational resources (raw materials, human resources (HR), capital, infrastructure, etc.) to maintain an organization's viability and existence in the long run and to provide it with sustainable competitive advantage and above-average returns (Ülgen & Mirze, 2004). Strategic management thus gains greater importance for organizations that compete in high-technology industries (e.g., IT, telecommunications, etc.) in an environment that can be characterized as fluid and unpredictable. Forming strategic alliances through licensing, joint ventures, and equity investments as well as engaging in mergers and acquisitions may constitute ways for companies to deal with this uncertainty, since such strategies might help them gain competitive powers through synergy creation and resource sharing. Thus, organizations may enhance their viability and survival in the changing environment.

In this chapter, a case study of the AVEA merger by two large GSM operators in the Turkish telecommunications sector is presented, first through a brief theoretical background description of high-technology industries, and then through the importance of strategic alliances and mergers within the telecommunications industry. We will then focus on some background information about the new entity, its principal shareholders, and on the Turkish GSM sector in general. These will be followed by the analysis of the merger process with special emphasis on IT, products, and HR. The study will end with concluding comments regarding the present position of the new entity in the market, the telecommunications sector in Turkey in general, and recommendations for further research.

Information and Communication Technologies and Strategic Partnerships

Since the early 1990s, the literature abounds with articles elaborating on information and communication technologies. Some of these articles focus on the role of IT in transforming the organizational structure into a flatter and more decentralized one with a faster decision-making process (Fulk & DeSanctis, 1995; Huber, 1990; Lucas & Baroudi, 1994). Others argue for the strategic role that IT is gaining to support and shape the business strategies (Henderson & Venkatraman, 1993). Some newer works also suggest that IT provides companies with a competitive edge through simplification, speed, and flexibility by incorporating IT into business practices (Ataay & Kalika, 2004; Powell & Dent-Micallef, 1997).

Today, the adoption and incorporation of IT into a company's operating system can practically be considered not a variable, but a constant. As a matter of fact, information and communication industries have grown so quickly they have created their own niche in the market, specializing in providing IT services to other industries as well as the general public. So, the rise in importance of network relationships at business-to-business (B2B) level (such as outsourcing IT-related services to specialized companies) is an obvious observation today. Moreover, one can easily notice various IT offerings by specialized companies at business-to-customer (B2C) level to elicit and meet demands from the public.

It is also clear that a technological convergence is going on in information and communication industries. Technological convergence refers to the ability of different network platforms to carry similar kinds of services that address the same market; or, in more concrete terms, creating a unique compatible consumer device that can function as telephone, television, and personal computer (Bores, Saurina, & Torres, 2003). Technological convergence, in that sense, changed the nature of competition in some industries by creating a greater technological and market uncertainty requiring huge investments. In this context, companies have faced and overcome these developments through the application of various strategies: they built standards for tempering technological uncertainty; they created demand with the supply of content for dealing with market uncertainty; and they tried to provide additional/complementary resources to make huge investments (Bores et al., 2003).

Finding complementary resources in order to make investments can be realized in several ways; forming strategic alliances is one of them (Yasuda, 2005). In the past, investments were mostly directed to developing internal know-how and new technology. However, firms gradually realized that in a continuously changing technological environment, it was no longer sufficient to engage only in internal development efforts (Verspagen & Duysters, 2004). In order to survive, organizations need to engage in exchange relationships with their environment and manage their dependencies in collaboration with other organizations upon scarce, valuable resources, as put forward by resource dependence theory (Pfeffer & Salancik, 1978, cited in Scott, 1992). Therefore, most firms seek to establish strategic alliances in order to increase their flexibility and gain access to another firm's technological resources. Moreover, it is also possible to enlarge this technological network, since through partnerships firms can gain indirect access to other firms' resources including technological relationships with different firms. Through such strategic partnerships it is also possible to gain access to others' overt and tacit resources and processes, which would not be otherwise available (Gürol, 2004; Neill, Pfeiffer, & Young-Ybarra, 2001).

These strategic partnerships—which help the firms to gain new competencies, preserve resources, share risks, move into new markets, and create opportunities for new investments (Hutt, Stafford, Walker, & Reingen, 2000)—evolve in time from strategic alliances to mergers and acquisitions. In fact, according to Utterback's model of technology life cycle, companies use alliances more often in their early stages, but mergers and acquisitions take place mostly after the transitional and maturity stages (Roberts & Liu, 2001). As knowledge is dispersed and disseminated over time, partnerships in the form of mergers become more practical and popular. When technology development requires pooling and using knowledge and technology that are controlled by two or more firms, this may increase the need for cooperation between complementary or even competitive actors (Amesse, Latour, Rebolledo, & Seguin-Dulude, 2004). Thus, "horizontal mergers between companies with complementary products and services" are one of the events that can happen in the maturity phase of the technological life cycle (Roberts & Liu, 2001, p. 29).

Mergers (forming a new entity by combining two or more companies) and acquisitions (buying a company for inclusion in the portfolio of the acquiring company) are the combinations that have the greatest implications over investment, control, integration requirements, and people management issues. Some of the reasons for mergers and acquisitions are (Schuler & Jackson, 2001):

- Market dominance, economies of scale;
- Spreading and sharing the risks, creating synergy;
- Growth for world-class leadership;
- Flexibility, leverage; and
- Talent gaining, knowledge and technology improvements.

The process can be analyzed in a three-stage model, as proposed by (Charman 1999, cited in Schuler & Jackson, 2001) and (Habeck et al. 1999, cited in Schuler & Jackson, 2001):

1. Precombination—planning and preparation;
2. Combination—integration of the partners; and
3. Solidification and advancement—the new entity.

Although mergers or acquisitions are the easiest way to grow, they may pose difficulties in terms of creating synergy and molding cultures. On the other hand, perseverance and diligence are necessary in the process and planning might increase the chances for success (Schuler & Jackson, 2001). Indeed, planning serves to assess the compatibility of partners' goals, determine the method of implementation, and indicate key challenges that might be encountered during the process (Mandal, Love, & Irani, 2003), considering, that the aforementioned facts might increase the success of mergers and acquisitions.

AVEA: Merger of Two GSM Operators in Turkey

In this part of the study, the merger formed by two GSM operators in the Turkish telecommunications sector will be reviewed. The newly established firm is named AVEA, which literally means A and A, with A standing for the initials of the two merging firms. After presenting the new firm briefly, some background information about the shareholders and the Turkish GSM sector in general, will be provided. Then the focus will be directed to AVEA, and the merger process will be analyzed with an emphasis on IT, network sections, products, and HR. The study will end with a discussion of AVEA's success and its recent position in the industry, general conclusions about the telecommunications sector, and technology environment in Turkey.

Method

The method adopted in this research is the case study. The case study method serves to investigate the holistic and meaningful characteristics of real-life events (Yin, 1989). With this method, it is possible to gain in-depth understanding of the factors that led to the merger decision and the way in which the process was realized. The method thus makes it possible to develop a basis upon which a theoretical framework can be constructed to explain this merger.

Researchers conducted semi-structured interviews with top managers in key positions related to IT and network planning in their offices at AVEA headquarters. The interview process took approximately one week, and a total of five managers were interviewed. The interview questions explored the merger process in general; technical adaptation problems; changes that occurred in the firm's technical capacity after the merger; products and services offered; new positioning of the merged firm in the market; cultural clash between the two firm's employees; and so forth. Some of the questions, especially those concerning financial data and economic gains, were left unanswered due to confidentiality issues. Interviews were partly tape-recorded and filed in notes. Each interview took approximately 30 minutes.

Researchers also benefited from the firm's official Web site (www.avea.com.tr), company-provided written materials, on-site observations, and small talk with other personnel during the interview period.

The Firm

TT&TIM Iletisim Hizmetleri A.S. was officially founded on February 19, 2004 through the merger of Türk Telekom's GSM operator Aycell and Is-TIM's GSM operator Aria. Is-TIM, was also a joint venture founded by Türkiye Is Bank (Is) and Telecom Italia Mobile (TIM). TIM and Türk Telekom equally own 40% stakes in the TT&TIM capital, while Is Bank Group holds 20% of the shares. During the short period just after the merger, pre-established brands of the participants Aycell and Aria were sustained under the umbrella of the new venture TT&TIM. As of June 23, 2004, the new AVEA brand was introduced to the market as a replacement of those two pre-established brands to announce the synergy gained after the merger. The commercial registered name of TT&TIM Iletisim Hizmetleri A.S. was changed to AVEA Iletisim Hizmetleri A.S. on October 15, 2004.

As stated during the interviews, the process of merging was managed by Aria's Italian team and an executive from Aycell; the whole process ended in one year. It was also stated that thanks to having a formerly state-owned enterprise (Aycell) as part of the merger, the regulatory process in the parliament concerning this transaction was somewhat expedited so as to reduce potential administrative problems likely to arise during the merger process.

The Shareholders

Türk Telekom, holding a 40% share in AVEA, is the leader in the Turkish telecommunications sector. Formerly a state-owned enterprise, it was privatized in 1994 and attained a more dynamic structure. With an established and well-managed telecommunication infrastructure throughout Turkey, Türk Telekom was the pioneer in the telecommunication sector with its automated switch phone capacity of over 21 million, 90.77% of digitalization ratio and three satellites, and ranked 13th among countries in terms of number of subscribers. Moreover, having introduced Turkey to GSM technology in the first months of 1994, Türk Telekom had laid the groundwork for mobile communication. After deregulation of license rights in 1998, Türk Telekom obtained the GSM 1800 Mobile Phone Operating License on March 16, 2000 and founded Aycell Communication and Marketing Services Inc. on January 8, 2001. Türk Telekom also introduced other technologies to the country. Recently, the company has been heavily investing in ADSL technology.

Telecom Italia Mobile (TIM), holding a 40% share in AVEA, is the leading European GSM operator in a single network and the leader in the Italian domestic market with approximately 27 million lines. Overall, the TIM Group with its subsidiaries and associate companies in Europe, South America, and the Mediterranean basin has reached an approximate number of 45 million lines. With a high level of penetration and rapidly evolving technology, TIM is considered an innovative company and a leader in traditional voice services as well as mobile phone services. It has opened Italy to the mass market and led the country into an advanced market position in Europe by introducing the family tariff in 1993. Later in 1996 the company introduced the market to prepaid telephone cards. More recently, it has introduced the concept of self-recharging. In June 2002 TIM introduced MMS to the country, with an open business model chosen for this interactive service supply based on the integration of a number of technological platforms (UNITIM, SimToolkit, WAP, and GPRS). The TIM Group is considered a prime technological and marketing partner partly because of its fixed price lists, innovative services, and operating processes (such as network, customer care, billing, and sales know-how) for its associates. Their flexibility and rapidity, economies of scale, and fast start-ups, put the company in a leading position in the world.

Türkiye Is Bank, holding a 20% share in AVEA, is the first and largest private bank in Turkey. It also has the leading role in Internet banking in the Turkish market. With a portfolio made up of public companies, Is Bank has the highest number of partners exceeding almost 250,000 shareholders. Being Turkey's largest private bank with its $22.3 billion of assets (as of Dec. 31, 2003), Is Bank provides services for industry, commerce, and the general public.

The Sector

The Turkish telecommunications market has a high growth potential given its low penetration rate (as reported by the chairman of the board of directors in an interview). The leading company and first mover to the GSM industry is Turkcell. Turkcell today possesses the largest customer base as well as the most advanced technology. Telsim, the second mover in the industry, serves the Turkcell-dominated market with alternative and attractive products. However, some governance problems in the group holding the majority shares of Telsim restrict its enlargement. But thanks to some of its customized offerings, such as low-priced packages provided for students, family, friend networks, and so forth, Telsim has built a significant amount of customers and ranks second after Turkcell. Aria and Aycell, which were founded consecutively, did not seem to have much share in the market due to their limited products and/or low-price packages unless they could make innovative and attractive moves in the industry. With the formation of AVEA, a breath of fresh air was introduced to the market; the new firm is heading toward creating brand-new product alternatives for customers and becoming a challenging competitor in the sector.

Management Structure

AVEA's board of directors is composed of Italians (40%) and Turks (60%). The chairman is an Italian, and the CEO is a Turk. The other top management positions, which include marketing and sales, network, information systems, finance, legal affairs, and HR, are equally divided among Italian and Turkish managers. The organization structure (divisions) is based on functional departments. The company headquarters is located in Istanbul.

Merging of IT and Network

The merger process took almost a year. In the planning and analysis phase, which could be considered one of the most important stages of the whole process, it was discovered that the two firms' systems were complementary in too many aspects ranging from customer regions to IT infrastructure.

It only took two months to analyze how to merge the IT infrastructure, and eight months to prepare strategies. In order to avoid any missing items, all strategies were written down even to their smallest components and planned on a step-by-step basis. In a sense, every potential inconvenience that could have happened in the IT merger was seriously assessed beforehand. With both having met few problems and having invested in technology, AVEA's IT infrastructure, today, is behind Turkcell's, but far above Telsim's.

The merging of IT infrastructures between the ex-companies was made by utilizing Ericsson's infrastructural systems, which were the common ground for both. The base stations had to be merged as well, leading to the closings and/or displacements of some stations. Since the firms independently covered almost all large cities in Turkey prior to the merger, some base stations had to be removed and relocated to other parts of the country. The decisions about which stations to remove were based on criteria such as the station's growth potential, rental fee (cheap or expensive), and its location (on ground or on tower). Today, the total number of base stations is 6,500, and the capacity has doubled.

International roaming contract numbers were 250 for Aria and 18 for Aycell. Aria's strength in this respect continued. However, it took some time to include AVEA in the roaming operators' list in place of Aria. This process is nearly complete, and there are now about 280 international roaming contracts.

The merger initially aimed to improve the quality of technical services and infrastructure, with no harm to customers. This objective was realized, and in that respect, managers and interested parties concurred that the AVEA merger, at least technically, was successful.

Merging of Target Population, and Products and Services

Before the merger, Aria expanded and positioned itself in the western territories of Turkey, while Aycell expanded in the eastern part. The merger, therefore, allowed the new company to cover and integrate all territories of Turkey. Having adopted a customer-oriented approach, AVEA aims at offering innovative and creative solutions to different groups of customers through reduced prices for students, state-employed officers, and prepaid or postpaid subscriptions. Furthermore, continuous promotion campaigns were launched to introduce such new offerings. Pre-merger products and services still continue for all ex-Aria and ex-Aycell customers, while the new product offerings are launched with AVEA brands solely available to AVEA SIM cardholders. However, if desired, these ex-firms' customers can easily replace their old SIM cards with new ones. In brief, as a brand-new alternative for the GSM sector in Turkey, AVEA aims at offering high technology with convenient prices. Indeed, it seems to be ranked second after Turkcell in regard to number of customers.

Merging of Human Resources and Organizational Culture

The qualified workforce of AVEA could easily find new jobs, and there were those who left the company of their own will. But this was not an expected outcome of the merger. The merger was realized with no waste of resources, including HR. However, since the ex-firm Aycell was a state-owned enterprise, changing habits and attitudes were considered likely to create some communication problems and misunderstandings in daily work practices. Training programs, although not extensive, were provided for employees to avoid the expected might-be outcomes and to enhance teamwork within the new company. However, managers today see such problems as minor issues and claim that the change in the organizational culture is occurring at a natural pace.

Discussion and Conclusions

The AVEA merger seems a clever strategic move taken by sides right in the midst of the technological convergence in the telecommunications sector. In an era in which multiple services can be provided by multiple channels of technology, AVEA unified power bases in order to reach a stronger competitive posture against Turkcell, the sector leader.

The two competing ex-firms Aria and Aycell had come to realize that they had complementary resources to offer each other and a merger would achieve a stronger market position. Technically, the merger was quite well planned, and the new infrastructure that was created was very powerful. Moreover, for strengthening the firm's market position against Turkcell, abundant and interesting promotional campaigns were launched in order to announce competitive offerings at lower prices and with more convenient conditions. So, the newly created AVEA was also successful marketing-wise in the competitive Turkish GSM industry.

The merger helped the two ex-firms gain access to each other's overt and tacit resources and use both firms' previous strategic ties with various organizations and stakeholders (e.g., international roaming operators of Aria being transferred to AVEA). The resource dependence approach explains this phenomenon. As mentioned previously, this approach claims that possessing a controlling power over environmental resources is a major challenge for all organizations to survive. After determining what significant and scarce resources they need, organizations search for ways to control these resources. They try to manage their dependencies by establishing a balance of power between themselves and their environment (Daft, 1998). In the AVEA case, a previous competitive relationship between two firms has turned into a symbiotic relationship through the merger, and thus it can be said that they created their own resources within themselves.

It is also possible to note the merging of the intellectual capital of the two firms. A synergy has been established between the firms with respect to structural capital, customer capital, and human capital (though to a lesser extent). The technological infrastructures of the firms have been well thought out and designed so that complementary points between the firms resulted in increased effectiveness. The customer capital has obviously increased by the summation of the two customer bases. Human capital, except for personnel in key positions and/or with key qualities, seems somewhat lost due to ineffective HR communication during the merger process. In this regard, the merging of the two cultures is left to proceed naturally. Although it is early to talk about a new overall culture (spirit of AVEA), more active and effective HR strategies and other activities to enhance ties among people and values could have been pursued in order to preserve valuable HR. Nevertheless, considering the GSM sector in Turkey and the gradual rise of AVEA to the second position after the leader (Turkcell), it is unlikely that the merger would be considered unsuccessful due to these softer reasons.

With regard to Utterback's technology life cycle model, the Turkish GSM sector is now entering the industrial maturity phase. As a relatively young market, the Turkish telecommunications sector has the possibility to grow. That is why the merger of the two ex-firms (and subsequent formation of AVEA) seems a good move to increase their joint competitive power in the face of future developments. The industry is ripe for further studies that could analyze developments and investigate how the GSM market expands with novel entrants in and outside the country.

References

Amesse, F., Latour, R., Rebolledo, C., & Seguin-Dulude, L. (2004). The telecommunications equipment industry in the 1990s: From alliances to mergers and acquisitions. *Technovation, 24,* 885-897.

Ataay, A., & Kalika, M. (2004). Perception du rôle stratégique des TIC & changement organisationnel: Etude France—Turquie. Retrieved from http://www.dauphine.fr/crepa/ArticleCahierRecherche/Conferences/agrh19mai2004/communication13.pdf

Bores, C., Saurina, C., & Torres, R. (2003). Technological convergence: A strategic perspective. *Technovation, 23,* 1-13.

Daft, R. L. (1998). *Organization theory and design* (6th ed.). Cincinnati: South-Western College Publishing.

Fulk, J., & DeSanctis, G. (1995). Electronic communication and changing organizational forms. *Organization Science, 6*(4), 337-349.

Gürol, Y. (2004). İşletme değerini arttırıcı bir unsur olarak entelektüel sermaye, bileşenleri, yönetimi ve ölçümü. *İ.Ü. İşletme Fakültesi Dergisi, 33*(2), 73-95.

Henderson, J. C., & Venkatraman, N. (1993). Strategic alignment: Leveraging information technology for transforming organizations. *IBM Systems Journal, 32*(1), 4-16.

Huber, G. P. (1990). A theory of the effects of advanced information technologies on organizational design, intelligence, and decision making. *Academy of Management Review, 15*(1), 47-71.

Hutt, M. D., Stafford, E. R., Walker, B. A., & Reingen, P. H. (2000). Case study: Defining the social network of a strategic alliance. *Sloan Management Review, 41*(2), 51-62.

Lucas, Jr., H. C., & Baroudi, J. (1994). The role of information technology in organization design. *Journal of Management Information Systems, 10*(4), 9-23.

Mandal, P., Love, P. E. D., & Irani, Z. (2003). Pre-alliance planning: Development of an information system infrastructure to support strategic alliance activities. *Management Decision, 41*(2), 132-140.

Neill, J. D., Pfeiffer, G. M., & Young-Ybarra, C. E. (2001). Technology R&D alliances and firm value. *Journal of High Technology Management Research, 12,* 227-237.

Powell, T. C., & Dent-Micallef, A. (1997). Information technology as competitive advantage: The role of human, business, and technology resources. *Strategic Management Journal, 18*(5), 375-405.

Roberts, E., & Liu, W. (2001). Ally or acquire? *Sloan Management Review, 43*(1), 26-34.

Schuler, R., & Jackson, S. (2001). HR issues and activities in mergers and acquisitions. *European Management Journal, 19*(3), 239-253.

Scott, W. R. (1992). *Organizations: Rational, natural, and open systems* (3rd ed.). NJ: Prentice Hall.

Ülgen, H., & Mirze, S. K. (2004). *İşletmelerde stratejik yönetim.* İstanbul: Literatür Yayınları.

Verspagen, B., & Duysters, G. (2004). The small world of strategic technology alliances. *Technovation, 24,* 563-571.

Yasuda, H. (2005). Formation of strategic alliances in high-technology industries: Comparative study of the resource-based theory and the transaction-cost theory. *Technovation, 25,* 763-770.

Yin, R. K. (1989). *Case study research: Design and methods.* Newbury Park: Sage Publications.

Chapter X

Product and Market Diversification and the Stock Market Response to Business Combinations: Empirical Evidence from the European Telecommunications Industry

Pablo Sánchez-Lorda, Universidad de Oviedo, Spain

Esteban García-Canal, Universidad de Oviedo, Spain

Abstract

In this chapter, we analyzed the stock market reaction to internationalization and diversification of the European telecom firms through acquisitions or strategic alliances. We analyzed these operations with the purpose of verifying to what extent the stock market reaction to these business combinations is influenced by information asymmetry, resource complementarity, and management costs. There are two main findings of this work. First, while acquisitions are more valued when entering into one of the more related businesses—telecommunication equipment shops—alliances, however, are only positively valued in the less related businesses.

Second, the strength of the competitive position of the firm moderates the relationship between product and market diversification and abnormal returns.

Introduction

There is a vast literature dealing with the stock market reaction to business combinations, that is, strategic alliances or acquisitions (García-Canal, Sánchez-Lorda, & Vidal, 2002). Although both alliances and acquisitions are means to expand firm boundaries into new markets or new countries, this literature is not of much help to identify criteria for choosing between these two entry modes. In fact, its results show that both alliances and acquisitions, aimed at expanding firm boundaries, could generate positive abnormal returns,[1] in other words, they could increase shareholders' wealth. The only existing comparative research that focuses on the stock market reaction in both types of business combinations is the work by Balakrishnan and Koza (1993). They argued and found empirical evidence that when the transaction of the external resource takes place in conditions of information asymmetry, joint ventures are more valuable than acquisitions since they protect the firm from adverse selection problems. However, not all empirical results support this view and some competing explanations—also with partial empirical support—exist. Resource-based theory suggests that when a company gains access to dissimilar resources through acquisitions, more synergies can be expected because the bidder can obtain synergies from the combination of resources that are not easily replicable by other bidders (Harrison, Hitt, Hoskisson, & Ireland, 1991, 2001). In the case of alliances, dissimilar resources open the door to more learning opportunities, but also to more conflicts of interests (Hamel, 1991). Finally, some researchers have highlighted the importance of management costs in organizational combinations (Hennart & Reddy, 1997; Kogut & Singh, 1988). Although potential synergies may exist, the key point to realize these synergies is the capability to fully integrate the target into the organizational structure of the bidder. When some factors exist that make such integrations difficult, this last perspective suggests that strategic alliances would be the preferred entry option.

Given the existence of competing explanations and the lack of conclusive empirical results, we present in this chapter some empirical evidence on the stock market reaction to the organizational combinations carried out by the European telecommunication operators. Ours is an empirical study in which we try to analyze to what extent the stock market reaction to business combinations is influenced by information asymmetry, resource complementarity, or management costs. The interest of focusing on the European telecommunications industry is twofold. On the one hand, it is a regulated industry in which there are some relevant entry barriers and, for this reason, alliances and acquisitions—instead of organic growth—are the main options to expand firm boundaries. On the other hand, this industry has suffered important environmental changes such as the technological development related to wireless communications and Internet as well as market deregulation, especially inside the European Union (EU). These changes have forced companies to quickly adapt to new technologies and to a wider geographical scope by means of gaining access to external resources—either through alliances or acquisitions—that seemed to be more relevant under the new environment.

This chapter is organized as follows. First, we summarize the results of previous empirical research analyzing the stock market reaction to alliances and acquisitions aimed at expanding firm boundaries. Then we show our results regarding the organizational combinations carried out by European telecommunications firms from 1986 to 2001. Finally, we outline the main implications of our results.

Literature Review

The aims of internationalizing and diversifying the firm have been two of the factors most analyzed in the literature about the stock market reaction to business combinations. As shown in the next paragraphs of this section, related and international acquisitions could be valued, positively, by the stock market.[2] Alliances whose activities are related to those of its own partners are also valued by the stock market,[3] and the same happens to international alliances.[4] Thus, previous research highlights the fact that external growth through alliances and acquisitions increases shareholders' wealth—although this evidence is not conclusive. As a consequence, this literature is not of much help to identify criteria for choosing between alliances and acquisitions.

In the following paragraphs we outline the main contributions of previous literature about each factor and each kind of operation. An initial analysis of these contributions allows us to see the lack of conclusive results for both factors, which opens several paths of research to advance the knowledge of this problem.

Expanding Boundaries through Internationalization: Previous Results[5]

Table 1 shows the average abnormal returns estimated in the studies that have analyzed the stock market reaction to strategic alliances and acquisitions aimed at expanding firm boundaries (international expansion and diversification). Table 2 reflects the results of these studies in more detail. Concretely, it shows if these studies have found a positive, negative, or non-significant influence of certain specific measures of internationalization and diversification over the abnormal returns obtained by the firms studied.

The results related to the internationalization of the companies through acquisitions show that international acquisitions create more value than domestic ones. In fact, most of the works about the acquisition of a foreign target have found a positive reaction of the stock markets. However, this positive effect seems to disappear as the cultural distance between the countries of origin and the destiny of the investment increases. In effect, almost every work about cultural distance has found a negative response of the stock markets—the cultural distance is measured using the index of Kogut and Singh (1988), based on the measures of Hofstede (1980). However, the work of Dewenter (1995) does not show significant results related to this issue, although this work is focused only in chemical and retail industries.

With regards to the aspects related to the internationalization by means of strategic alliances, this factor has practically not been contemplated per se. Only two works (Finnerty, Owers, & Rogers, 1986; Park & Kim, 1997) have

Table 1. Abnormal returns associated to acquisitions and alliances aimed at expanding firm boundaries

Operation	Factor studied	Study	Period of study	Sample size	Measure of return	CAR†
Acquisitions	Internationalization	Doukas & Travlos (1988)	1975-1983	301	AR (0)	- 0,08
		Chatterjee, Lubatkin, Schweiger, & Weber (1992)	1985-1987	52	CAR (- 10,5) CAR (- 3,3)	- 3*** - 6**
		Cakici, Hessel, & Tandon (1996)	1983-1992	195	CAR (0,1)	0.63***
		Frame & Lastrapes (1998)	1990-1993	54	CAR (- 5,5)	- 0,30**
		Rad & van Beek (1999)	1989-1996	56	CAR (- 1,0)	- 0,25
		Brooks, Feils, & Sahoo, (2000)	1981-1989	272	CAR (- 1,0)	- 0,36**// - 0,31[b]
		Eckbo & Thorburn (2000)	1964-1983	1261	CAR (- 12,- 1) Monthly data	2,37***
		Pantzalis & Kohers (2000)	1992-1997	151	CAR (0,1)	1,93***
	Diversification	Lubatkin (1987)	1948-1979	315	CAR (- 18,- 1) CAR (1,18)	8,4*** - 0,5
		Singh & Montgomery (1987)	1975-1980	77	CAR (-5,25)	- 0,6//- 1,9[a]
		Datta & Puia (1995)	1978-1990	112	CAR (- 1,0) CAR (- 5,5)	- 0,42* - 0,72*
		Hubbard & Palia (1999)	1961-1970	392	CAR (- 5,5)	1,62***//0,24[a]
		Leeth & Borg (2000)	1919-1930	466	CAR (- 1,0) Monthly data	2,43***
		Shelton (2000)	1962-1983	199	CAR (- 1,0)	- 0,6**
		DeLong (2001)	1988-1995	280	CAR (- 10,1)	- 1,68***

Table 1. (continued)

Operation	Factor studied	Study	Period of study	Sample size	Measure of return	CAR†
Alliances	Internationalization	Finnerty, Owers, & Rogers (1986)	1976-1979	190 jvs	AR (0)	0.22 //0.19 [b]
		Lummer & McConnell (1990)	1971-1980	416 jvs	CAR (- 1,0)	0.4**
		Chen, Hu, & Shieh, (1991)	1979-1990	88 jvs	CAR (- 1,0)	0.71***
		Crutchley, Guo, & Hansen, (1991)	1979-1987	146 jvs	CAR (0,1)	0.99***
		Chung, Koford, & Lee, (1993)	1969-1989	230 jvs	AR (0)	- 0.16
		Waheed & Mathur (1995)	1963-1989	35 jvs	CAR (- 1,0)	- 0.11**
		Ojah, Seitz, & Rawashdeh, (1997)	1980-1991	125 jvs	AR (0)	0.021
		Park & Kim (1997)	1979-1988	158 jvs	AR (0)	0.763***
		Chen, Ho, Lee, & Yeo (2000)	1979-1993	174 jvs	CAR (- 1,0)	0.96***
		Merchant & Schendel (2000)	1986-1990	101 jvs	AR (0)	- 0.2
		Reuer & Koza (2000)	1985-1995	297 jvs	CAR (- 1,1)	0.655** // 0.177 [b]
		Vidal (2000)	1987-1997	72 global alliances	AR (0)	0.19**
	Diversifi-cation	Koh & Venkatraman (1991)	1972-1986	175 jvs	CAR (- 1,0)	0.87***
		Balakrishnan & Koza (1993)	1974-1977	64 jvs	AR (0). Monthly data	1.19***
		Mohanram & Nanda (1996)	1986-1993	253 jvs	CAR (- 1,1)	0.19
		Chan, Kensinger, Keown, & Martin, (1997)	1983-1992	345 all.	AR (0)	0.64***
		Fröhls, Keown, McNabb, & Martin (1998)	1987-1992	320 jvs	CAR (0,1)	0.89***
		Wu & Wei (1998)	1985-1992	209 all.	AR (0)	0.89**

† CAR: Cumulative Abnormal Return in a given period (%).
[a] Related//Unrelated acquisitions. [b] Domestic//International acquisitions.
* p < .1 ** p < .05 *** p <.01

Table 2. Determinant factors of value creation in acquisitions and alliances

Op	Factors		Positive Reaction	Non-significant effect	Negative Reaction
ACQUISITIONS	Internationalization	Foreign Target	Doukas & Travlos (1988); Cakici, Hessel, & Tandon (1996); Frame & Lastrapes (1998); Delios & Beamish (1999); Brooks, Feils, & Sahoo (2000); Eckbo & Thorburn (2000); Pantzalis & Kohers (2000); DeLong (2001)	Rad & Beek (1999)	
		Cultural Distance (Kogut & Singh, 1988)		Dewenter (1995)	Chatterjee, Lubatkin, Schweiger, & Weber (1992); Datta & Puia (1995); Very, Lubatkin, Calori, & Veiga, (1997)
	Diversification	Target from other industry	Hubbard & Palia (1999); Leeth & Borg (2000)	Doukas & Travlos (1988)	Morck, Shleifer, & Vishny (1990); Shelton (2000)
		Degree of relation among businesses	Balakrishnan & Koza (1993); Singh & Montgomery (1987); Datta & Puia (1995); Shelton (2000); DeLong (2001)	Lubatkin (1987); Seth (1990)	Chatterjee (1986)

Table 2. (continued)

Op		Factors	Positive Reaction	Non-significant effect	Negative Reaction
ALLIANCES	Internationalization	Foreign partner		Finnerty, Owers, & Rogers (1986); Park & Kim (1997)	
		Cultural Distance	Kim & Park (2002)	Merchant & Schendel (2000); Vidal & García-Canal (2003)	
	Diversification	Similarity among partners' businesses	Koh & Venkatraman (1991); Chan et al. (1997); Park & Kim (1997); Wu & Wei (1998); Vidal & García-Canal (2003)	Merchant & Schendel (2000)	Balakrishnan & Koza (1993); Mohanram & Nanda (1996); Reuer & Koza (2000)
		Similarity among partners' and alliance's businesses	Koh & Venkatraman (1991); Merchant & Schendel (2000)	Fröhls, Keown, McNabb & Martin (1998)	

considered whether international or domestic character of the alliances could be a determining factor of the value creation. Besides, neither of these works reaches significant conclusions. It does not mean, however, that the event studies have not taken care of the analysis of the international alliances. On the contrary, a great part of these studies, whose results have been summarized in Tables 1 and 2, have analyzed joint ventures. These works show that, in general, the market reacts positively to the announcement of an alliance to internationalize. Yet, it is possible to emphasize that it is, in fact, in this type of alliances where negative responses of the market are seen. With the aim of identifying and explaining in what circumstances the international alliances can create or destroy value, it has been examined if the distance between the national cultures of the partners—measured again through the index of Kogut and Singh (1988)—is a determining factor of the stock market reaction to alliance formation. In this case, the obtained results are less conclusive than in the case of acquisitions. Cultural distance has not been influencing—at least, in a negative form—the abnormal returns associated to the formation of strategic alliances. It may be because, in culturally distant countries, alliances are a more attractive means of international expansion than acquisitions (Kogut & Singh, 1988).

Summing up, international acquisitions seem to be better valued by the stock markets than international alliances; although, the cultural distance among the firms involved in the operation seems to be more harmful to bidders of international acquisitions than to firms that internationalize by means of alliances. This can be a confirmation of the management cost approach. However, there is an opportunity for further research by jointly analyzing acquisitions and alliances in order to clarify the real effect of internationalization on the shareholders of the firms involved.

Expanding Boundaries through Diversification: Previous Results

In relation to diversification, as shown in Tables 1 and 2, this is the line of research that remains most open within the field of acquisitions, mainly, the relative to the main effect of diversification, for which contradictory results exist. In fact, a wider debate in the literature on corporative finance and strategy exists about whether diversification *per se*—independently of the form in which takes place—creates value or not. Nowadays, works are appearing that demonstrate how diversification by itself does not destroy value. Among these works, Campa and Kedia (2002) and Villalonga (2004) deserve special mention. Both works find evidence that the diversification by itself does not destroy value. The companies that decide to diversify their activities have characteristics that may explain most of the negative effects that this type of acquisitions can have on the stock market response. An opportunity of research in the field of acquisitions, therefore, is to analyze the effect of diversification on the abnormal return associated to these operations by incorporating variables that control such characteristics of the firms.

The influence of the type of diversification seems to be clearer. The results reached in most of the previous research point out that, the more diversification is valued by the markets the greater the degree of relation between the businesses of the bidder and the target is. Among the most relevant works, we could mention Singh and Montgomery (1987) and Shelton (2000). However, Chatterjee (1986) concludes the opposite. On the other hand, Lubatkin (1987) determines that the market does not value, in a different way, the related or not related acquisitions. Trying to clarify the effect of this factor, Seth (1990) postulates that, in related as well as in unrelated acquisitions, there exists value creation due to existing synergies between the companies. The difference is based on the source of these synergies in both types of acquisitions. Whereas, in related acquisitions these synergies are mainly associated to the relative size of the company target and the bidder; in unrelated ones, they are not linked to the possibility of saving taxes by way of the combination of both companies' debt.

Another idea pointed out in this work, is that the mechanisms of classification of the acquisitions must be sharpened because it seems that this can be the cause of previous discord in this subject. Later on, Flanagan (1996) retakes the argument of Seth (1990) and defines two types of acquisitions—purely related and purely not related—based on the strategic similarity of firms. The conclusion of this work is that the companies that acquire purely related ones obtain benefits higher than those that acquire non-related companies. Other authors such as DeLong (2001) continue analyzing what happens with the degree of relation and the returns of the companies, although this one is a debate that still remains open.

All these results have to be taken with care when comparing them, because, as said before, the variables that measure relatedness are heterogeneous. While some use dummy variables that take value 0 if the acquisition is not related and 1 otherwise, others as Balakrishnan and Koza (1993) measure the distance that separates the activities of the companies using continuous variables and standard industrial classification (SIC) codes. For that reason, it seems that to get to define a more refined variable than the one used to measure that degree of relation can be an objective to reach in future works. Also, it would be interesting in this field to introduce controls of characteristics of the companies that could influence the type of diversification that is made by means of acquisition—as it was previously considered for the main effect of the diversification—to measure its influence better.

In the case of the alliances, the empirical evidence around the diversification is also mixed. We can identify two main groups in this field. On the one hand, the works that analyze alliances between companies in the same industry. On the other hand, those that study the influence between the degree of relation of the businesses of the partners and the activities they develop in collaboration. The alliances between competitors seem to create more value than those created between partners who belong to different industries, as is shown in Table 2. From the point of view of resources, alliances between competitors would allow obtaining greater synergies and could benefit the partners with a greater market. Yet, all the obtained evidence on the matter is not positive. Thus, from the information asymmetry perspective, alliances between related partners create less value because information asymmetry does not exist. This impedes a correct evaluation of the capabilities contributed by each partner, which would therefore, make acquisitions a more attractive option. On the other hand, and independently of who is the partner, the market seems to value more positively, alliances that have a greater relation to the main businesses of the partners. As shown in Tables 1 and 2, almost all the obtained evidence on this issue is positive and significant. Moreover, some works (Koh & Venkatraman, 1991) demonstrate that the partner who obtains greater value creation with the alliance is the one whose activities are related to those made in collaboration.

Summing up, in terms of diversification the evidence about the effect on the abnormal returns of acquisitions and alliances is mixed. What is more clear is that the higher the relation between the businesses of the bidder and the target, or those of the partners that form an alliance and the one of the alliance, the better. Nevertheless, much work remains to be done related to the real effect of diversification *per se* over the performance of acquisitions and alliances. Particularly, studies that clarify the influence of diversification—and also internationalizaion—are also necessary, especially those studies controlling by all relevant variables—something that has not always been done.

Methods

Sample

In order to test how the market values diversification and internationalization through acquisitions and alliances, an initial sample including all of the acquisitions and alliances carried out by European telecommunication companies between 1986 and 2001 was obtained from the Secuities Data Company (SDC) database. This database is the most reliable source for identifying these operations and has been widely used in the fields of strategy, management, and finance. We focused specifically on the providers of telecommunication services, that is, those firms whose main SIC code is 4812 (Radiotelephone Communications) or 4813 (Telephone Communications, Except Radiotelephone).[6] A total of 643 acquisitions and 830 alliances for the studied period were identified. Our goal was to analyze how the formation of these business combinations affected the share price of a European telecom company. We have treated each abnormal return as a separate case; we have calculated the abnormal return for the bidder firm in the case of acquisitions; and for all the European partners in the case of strategic alliances. Because of the methodology used, each combination announcement had to meet the following criteria so as to be included in the final sample:

1. The European firms' daily stock prices were available in the DataStream database.[7]
2. The precise date of the announcement could be identified in the Lexis-Nexis database.[8]
3. No major confounding announcement that could contaminate the effect of the studied event was made within a ± five-natural day period around the announcement day.[9]

After the elimination of all the events that did not fit the criteria mentioned previously, the final sample consisted of 271 and 372 experiences of European telecom firms in acquisitions and in strategic alliances, respectively, carried out by 40 firms.

Dependent Variable and Method of Analysis

Standard event study methodology was used to examine if the number of firms that had previously expanded their boundaries to a specific field, through acquisitions or alliances, influences the abnormal returns generated by these operations.

As dependent variable, we used the cumulative abnormal returns between ± three trading days around the public announcement of the business combination. We used this period because it is the widest window in which we could guarantee the non-contamination of our events, due to our five-day screening for confounding events. We followed Brown and Warner's procedure (1985) in order to determine the reaction in the stock price of the financial assets under the announcement of certain relevant events. Abnormal returns are defined as the difference between actual returns and those returns that should be expected, according to a market model (Sharpe, 1964).[10] The estimation of the market model was carried out over a 180-day period beginning 200 days before the date of the announcement (t = - 200) and finishing 21 days before this same date (t = - 21), t = 0 being the announcement date. We excluded the 20 days prior to the announcement from the estimation of the market model so as to remove data that might be affected by the event. Their inclusion might have led to an undervaluation of the abnormal returns, since the effect of the announcement would have partially been incorporated in the expected returns.

Table 3 shows the results of the event study. For the whole sample, we observed that the companies that carry out business combinations obtained on average, negative and significant returns—except for the interval of accumulation [-1,1]—that oscillated between - 0.02% and - 0.32%. In acquisitions, we saw how the bidders obtained negative abnormal returns on average, when carrying out this type of operations. However, in the case of the alliances, although returns obtained were on average, negative in almost all the intervals, these abnormal returns were only statistically significant for the interval [- 2,2], in which a loss of value of - 0.28% is shown with a significance level of 90%.

As for the percentage of positive events, we observed that between 42% and 48% of the firms in all samples obtained positive abnormal returns. These high percentages show that strategic alliances and acquisitions were sometimes beneficial and sometimes disadvantageous for firms.

Dividing the samples attending to the geographical scope of the combination, Table 4 shows that the worst operations seem to be domestic acquisitions. In every interval of accumulation, these operations obtain, on average, negative and

Table 3. Abnormal returns (%)

Interval	AR whole sample	% Pos.	AR acquisitions	% Pos.	AR alliances	% Pos.
[- 3,3]	- 0.32**	45	- 0.31**	43	- 0.34	47
[- 2,2]	- 0.22***	45	- 0.16**	42	- 0.28*	46
[- 1,1]	- 0.04	48	- 0.02**	47	- 0.06	48
day 0	- 0.02*	46	- 0.03	45	- 0.02	47
N	643		271		372	
		* p < .1	** p < .05	*** p <.01		

Table 4. Abnormal returns associated to domestic and international combinations (%)

Interval	Domestic		International	
	Acquisitions	Alliances	Acquisitions	Alliances
[- 3,3]	- 0.53*	- 0.33	- 0.17	- 0.35
[- 2,2]	- 0.16**	- 0.03	- 0.18	- 0.35**
[- 1,1]	- 0.27**	- 0.20	- 0.14	- 0.02
day 0	- 0.06	0.12	- 0.08	- 0.06
N	104	84	167	288
	* p < .1	** p < .05	*** p <.01	

significant abnormal returns. In the case of alliances, only those involving partners from different countries seem to be harmful for firms' shareholders, although this result is only significant for the interval [- 2,2].

Finally, Table 5 shows the results related to operations aimed at diversifying businesses or not, either through acquisitions or alliances. The clearest result presented in the table is that acquisitions used for diversification are negatively valued by the stock market—regardless the interval of accumulation used. In the case of alliances, these operations are also negatively valued by the stock market when they imply entering into other businesses, although this result is only

Table 5. Abnormal returns associated to diversifying and non-diversifying combinations (%)

Interval	Diversifying		Non-diversifying	
	Acquisitions	Alliances	Acquisitions	Alliances
[- 3,3]	- 1.26***	- 0.20	- 0.61	- 1.46
[- 2,2]	- 0.79***	- 0.27**	- 0.46	- 0.37
[- 1,1]	- 0.63***	- 0.05	0.58	- 0.20
day 0	- 0.28	.04	0.22	- 0.49
N	133	329	138	43

* p < .1 ** p < .05 *** p <.01

significant for the interval [- 2,2], and the effect is not as high as that of acquisitions (- 0.27% in alliances vs. - 0.79% in acquisitions).

Variables Related to Resource Complementarity, Information Asymmetry, and Management Costs

In order to measure the existence of resource complementarity, information asymmetry, or management costs, we focused our attention in the internationalization and diversification of telecom firms, given that in processes of internationalization and/or diversification, all those factors—resource complementarity, information asymmetry and management costs—are higher. Concretely, we used the variables described below. First, we measured diversification through the following dummy variables:

- **Market Diversification:** This is a dummy variable, valued 1 when the firm is moving outside its home country and 0 otherwise.
- **Product Diversification:** This is a dummy variable, valued 1 when the firm is to enter into an industry different from its core business (SIC's 4812 and 4813) and 0 otherwise.
- **Market and Product Diversification:** This is a dummy variable, valued 1 when the firm is to enter both into another country or geographical area and into an industry different from its core business and 0 otherwise. This variable allowed us to isolate the maximum diversification.

These three variables were created based on the information provided by SDC. This database facilitates information about the geographical scope of acquisitions and alliances—the country of the target company and the geographical area towards which the alliance goes—and also provides information about the business scope of the combinations—the SIC code of the target firm and of the alliance.

In the case of alliances, as previously mentioned, previous works have also taken into account the nationality and the industry of the partners—same or different. On the one hand, international alliances imply higher management costs, information asymmetry, and resource complementarity. On the other hand, alliances among competitors involve higher management costs related to avoiding opportunistic behavior. Thus, in order to control these factors we also created the dummy variables *International Alliance* and *Competitors*. The first one is valued 1 when the partners of the alliance belong to different countries and 0 otherwise. The second variable is valued 1 when all partners belong to the telecom industry and 0 otherwise.

As *Tobin's Q* is a variable commonly used in many event studies related to acquisitions and alliances, we created the variable *Tobin's Q*[11] in order to approximate the strength of the market position of the firms. As suggested by Chung and Pruitt (1994), this variable was estimated as the ratio of the market value of equity plus the book value of debt to the book value of total assets. This variable was calculated on December 31st of the year immediately previous to the acquisition or the alliance because the data needed to estimate it—collected from DataStream—is only annual. We also multiplied *Tobin's Q* by the variables previously defined so as to see whether the strength of the market position of each firm moderates the effect of product and/or market diversification or not.

Alternative measures of market diversification were created. Concretely, as a refinement of the dummies described previously, we considered if the geographical region, entered through acquisitions or alliances, could affect the impact of such operations over the stock market response. In order to achieve this objective, we constructed four dummy variables following the procedure of López-Duarte and García-Canal (2002), valued 1 when the operation supposed the entry in the EU, Latin America, Organization for Economic Cooperation and Development (OECD) countries or other countries—EU, Latin America, OECD and other countries, respectively—valued 0 otherwise. The *Cultural Distance* of the firms involved in the operation was also measured. We created this variable by means of the Kogut and Singh index (1988), using Hofstede's revised measures (2001) as input. In the case of alliances, when they involved more than two partners, we followed the procedure employed by Kim and Park (2002). Specifically, for each pair of partners we calculated the Kogut and Singh (1988) index and, afterward, the average between these indexes.

Another alternative measure of market diversification is the variable *Product Distance*. We created this variable following the procedure of Balakrishnan and Koza (1993). Concretely, this measure is based in the three-digit SICs of the firms involved in the business combination and, in the case of alliances, also the SIC of the alliance. According to their methodology, *Product Distance* was calculated in the following way:

$$\textit{Product Distance}_{1,2} = |SIC_1 - SIC_2| \ / \ \text{max. distance},$$

being:

SIC_1 = the three-digit SIC of the firm that expands its boundaries.

SIC_2 = the SIC of the target firm, in the case of acquisitions, and the SIC of the alliance, in the case of this type of combinations.

max. distance = maximum possible distance in each subsample (393 in acquisitions and 476 in alliances).

Again, we multiplied this variable by *Tobin's Q* with the same purpose as before.

Finally, the maximum desegregation of the market diversification could be to consider the different industries in which a firm enters. According to this, we have classified our sample of alliances and acquisitions in the following fields: telecommunications (fixed and wireless); distribution (shops); TV and entertainment; directories; telephone equipment; hardware; software; communication infrastructures (cables and networks); and Internet service providers. These groups were formed after a detailed analysis of the patterns of expansion of these companies as well as by analyzing specialized literature. They comprise more than 85% of the total number of cases studied. The remaining 15% of cases were diversifying movements toward other industries that were directly identified by their four-digit SIC codes.

All these variables allowed us to approximate the influence of information asymmetry, resource complementarity, or management costs on the abnormal returns associated to acquisitions and alliances. We expected different effects of carrying out such operations over the stock market response to them, depending on the theoretical approach used to explain these effects.

According to the information asymmetry perspective (Balakrishnan & Koza, 1993), we expected a negative effect of diversification on the abnormal returns of acquisitions and positive in the case of alliances. In fact, in scenarios with high information asymmetry—like when a firm is entering into a new industry—one of the main problems that firms may face is the valuation of the resources they are trying to gain access to. The firms should then use more flexible means to enter into these scenarios, in order to avoid opportunistic behavior. Thus, alliances should be better valued than acquisitions under these circumstances, as the former imply less commitment of resources and are more easily broken than the latter.

According to the resource complementarity approach (Harrison et al., 1991, 2001), the expected influence of resource complementarity is positive, regardless of the kind of operation. This line of research postulates that the reason that justifies business combinations is the taking advantage of the existing complementarities—synergies—among firms' resources. Thus, acquisitions or alliances that gain access to complementary resources have the same benefits: both of them exploit the potential synergies arising from the combination of these resources.

According to the management cost perspective (Hennart & Reddy, 1997; Kogut & Singh, 1988), there is not any clear effect of acquisitions or alliances depending on their management costs. The key point of this perspective is to be able to fully integrate the target or the partner into the organizational structure of the bidder, and it is not clear when acquisitions or alliances are better means to achieve this objective. In fact, when the cultural distance of the firms involved in a combination is high, there should be a problem of digestibility that should increase management costs. Thus, alliances should be a better means to expand firm's boundaries under this situation.

Control Variables

To test the robustness of our results, we also included several control variables in our estimations. Firstly, year and company dummies were introduced with the aim of controlling the possible influence of time and any unobserved heterogeneity inherent to each company on the abnormal returns.

Secondly, the following variables were introduced in order to control the influence that the specific characteristics of the business combination, studied in previous literature, could exert over the abnormal returns.

GDP growth: This variable measures the growth experienced by the gross domestic product (GDP) of the geographical area entered through the operation in the year prior to it. The data needed to create this variable were obtained from the OECD statistics and from Eurostat. Hennart and Park (1993) show that the growth rate in the geographical area towards which a company goes, conditions the choice between greenfield investments and acquisitions. Thus, this variable is included in order to control this effect.

With regard to the control variables exclusively used for acquisitions, we measured the level of commitment of the acquisitions by means of these three dummies: Partial Acquisition, valued 1 if it is a first acquisition of less than 100% of the capital of the target company, *Total Acquisition*, valued 1 if it is the purchase of 100% without previous presence in the target capital, and *Accumulation*, valued 1 if it is not the first purchase made by the bidder over the target firm's shares. We excluded the first dummy variable to avoid perfect correlation problems. These variables were included in order to control the influence that the percentage of capital acquired and the previous presence of the bidder in the target's capital could exert on the abnormal returns associated to each acquisition. In fact, some works like Bradley, Desai, and Kim (1988) or Kuipers, Miller, and Patel (2003) have taken these facts into account.

For the subsample of alliances, we introduced the *R&D, Manufacturing,* and *Marketing* variables, which are non-excluding categories that allowed us to control the functional roles of the alliance. Many previous studies have examined how the alliance functional role, especially R&D alliances (Das, Sen, & Sengupta, 1998; Merchant & Schendel, 2000), could affect the abnormal returns of the firm's shareholders.

The last control variables introduced are *Partners*, which control the number of partners that form each alliance, and *JV*, a dummy variable that takes the value of 1 if the alliance has given rise to the creation of a joint venture, and 0 otherwise. The first variable was introduced as several works highlighted the importance of the number of partners and its effect on the performance of alliances.[12] The inclusion of the variable *JV* is very important, as one of the main classifications in the literature about alliances, it divides them in strategic alliances and joint ventures (Gulati, 1995).

With all these variables, we estimated several multiple linear regression models for the subsamples of acquisitions and alliances. Specifically, we estimated three models for each subsample. The first ones contain the diversification dummies—Models 1a and 1b. Models 2a and 2b contain geographical dummies, *Cultural Distance* and *Product Distance*. The last models—Models 3a and 3b—also consider the industry dummies.

Results

Table 6 presents the results of the three multiple linear regression models estimated in the subsamples of acquisitions and alliances—Models 1a, 2a, 3a and 1b, 2b, and 3b. Each model includes the value of the coefficients of the independent variables and an indication of their significance level. The explana-

Table 6. Multiple linear regression models [a]

Independent Variables	Acquisitions			Alliances		
	Mod 1a	Mod 2a	Mod 3a	Mod 1b	Mod 2b	Mod 3b
Year dummies	Included	Included	Included	Included	Included	Included
Company dummies	Included	Included	Included	Included	Included	Included
Intercept	103.2*	- 50.9	- 0.28	- 12.30	- 24.52	- 23.76
Control Variables:						
GDP Growth	- 0.27*	- 0.31*	- 0.30*	- 0.23	- 0.04	- 0.05
Total Acquisition	- 0.30	- 0.24	- 0.94			
Accumulation	- 0.79	- 0.56	- 1.10			
Joint Venture				0.07	0.01	- 0.10
R&D				0.65	0.80	0.65
Manufacturing				- 0.41	- 0.29	- 0.53
Marketing				0.09	- 0.03	0.33
Number Of Partners				- 0.21*	- 0.27**	- 0.27***
Variables related to resource complementarity, information asymmetry, and management costs:						
Market Diversification	- 1.54			0.29		
Product Diversification	- 0.84			- 2.14		
Market & Product Div.	1.41			- 2.22		
Tobin's Q	- 0.15	0.74	0.66	- 1.51	1.64	1.55
Q X Market Diversification	0.60			0.06		
Q X Product Diversification	- 0.22			2.35		
Q X Market & Product Div.	- 2.42*			1.36		
Competitors				2.68		
International Alliance				- 3.69*		
Q X Competitors				- 1.70		
Q X International Alliance				2.03*		
European Union		- 1.18	- 1.07		- 0.27	- 0.10
Latin America		0.47	1.30		2.14	2.55*
OECD		- 0.15	- 0.06		- 2.38***	- 2.22***
Other Countries		- 1.13	- 1.15		0.07	0.05
Cultural Distance		- 0.24	- 0.30		- 0.20	- 0.18
Product Distance		- 8.15***			2.08	
Q X Product Distance		2.35*			0.11	
Equipment			- 1.52			0.45
Shops			3.09*			- 1.83
TV			- 2.77			1.80
Directories			- 4.58**			
Hardware			- 6.52***			- 0.34
Software			- 4.59**			1.34
ISP			- 1.78			0.59
Cables						- 0.98
Networks						- 0.02
Other Industries			-0.91			1.63*
N	209	209	209	319	319	319
Model F	1.93***	1.84***	1.65**	1.39*	1.38*	1.32*
Adjusted R Squared	17.47%	16.99%	14.81%	5.75%	5.30%	4.89%

[a] Dependent variable: CAR (-3,3) * p < 0.1; ** p < 0.05; *** p < 0.01

tory power of all the models—measured via the F-statistic—is statistically significant.

Focusing our attention on the diversification dummies, we can say that in the field of telecommunications, the stock market does not value more the alliances or acquisitions aimed at expanding firm boundaries than those carried out in their home country and in their core business. The only situation in which the market reacts differently is that of maximum diversification and for those firms with a high Tobin's Q. In the telecom industry, Tobin's Q is an indicator of the strength of the market position of the firm in its current markets. As the coefficient of the variable *Q x Market and Product Diversification* shows, these firms—those with a high Tobin's Q—are penalized by the stock markets when they move outside their core businesses abroad, using acquisitions. Thus, it will be better for the firms with a high Tobin's Q to exploit their own resources in the home market.

In relation to alliances, international alliances carried out by firms with a high Tobin's Q are more valued by stock markets than alliances with partners of the same country, as observed in Model 1b. Concretely, the net effect of the formation of international alliances, which depends on the value of the Tobin's Q, will be positive with Q's greater than 1.82 and this value is approximately the average Q of our sample.

Another important result is that acquisitions could be seen as a means to leverage the competitive position in the home market expanding to unrelated industries, as the variable *Q x Product Distance* shows in the case of acquisitions. However, this result does not appear in the case of alliances. What we can see in these operations is that entering in OECD countries is penalized by the stock markets, whereas, entering into Latin America is positively valued—at least in Model 3b. These results could be explained if we have in mind the risk associated with entering into such geographical fields and the advantages of alliances related to this risk. In OECD countries the political stability is higher and the risk associated to an opportunistic behavior of the government is lower. An alliance allows the firm to avoid the risk associated to the uncertainty about the behavior of the government of the host country. In fact, in these agreements, it is easier to turn back to the initial situation of separated firms than in acquisitions, so it seems that alliances allow firms to cope better with this risk.[13] In OECD countries the problem of risk is not important, and then alliances are not needed to enter into these countries. On the other hand, Latin American countries have a higher political risk, so alliances are a more valuable means to enter into such countries.

Finally, acquisitions are more valued to enter into related markets to reinforce the competitive position of the firm, that is, expanding activities to telecommunication equipment shops. Alliances, however, are better means to enter into distant businesses, as the coefficient of the variable *Other Industries* shows in Model 3.

With regards to control variables, the most remarkable result is the effect of a higher number of partners over the abnormal returns of alliances. As previous works pointed out, a higher number of partners augment coordination and motivation costs (García-Canal, 1996; Oxley, 1997). Each new partner requires additional effort so the total amount of investments that are needed to make the alliance work is higher (García-Canal, Valdés-Llaneza, & Ariño, 2003). As a consequence, partners have fewer incentives to invest in the relationship, which aggravates the functioning of the alliance. Additionally, a higher number of partners makes the development of the agreement more difficult. A new partner not only increases the required amount of relational investments, it also reduces the chances to profit from these investments, given that it is more difficult to define new projects that meet the requirements of each partner. A final factor that limits the value that can be created in these alliances is that each new partner makes more difficult to put into practice the reciprocity mechanism (Parkhe, 1993). Having said that, the negative effect of the number of partners on the abnormal returns of alliances is not surprising.

Discussion and Conclusions

In this chapter, we analyzed the stock market reaction to internationalization and diversification of the European telecom firms through acquisitions or strategic alliances. We analyzed these operations with the purpose of verifying to what extent the stock market reaction to these business combinations is influenced by information asymmetry, resource complementarity, and management costs. We expected that, if the best explanation of stock market response were information asymmetry, then alliances should be a more valuable means to expand boundaries than acquisitions, which should be negatively valued by stock markets the higher the asymmetry (Balakrishnan & Koza, 1993). If the best explanation were the complementarity of resources, the results of acquisitions and alliances aimed at expanding firm boundaries should be positive in both cases. Finally, if the best argument were that provided by the management costs approach (Hennart & Reddy, 1997; Kogut & Singh, 1988), cultural distance and other factors associated to the business combination would increase alliance benefits versus acquisitions, although other factors, as previous experience in integrating the target firm of an acquisition in the organizational structure of the bidder—that could reduce the digestibility problem—could increase the benefits of using an acquisition to gain access to the resources of the target.

There are two main findings of this work. First, while acquisitions are more valued when entering into one of the more related businesses—telecommunication equipment shops—alliances, however, are only positively valued in the less related businesses. Second, the strength of the competitive position of the firm moderates the relationship between product and market diversification and abnormal returns. Firms with high q values are more valued when making acquisitions in their home country and when expanding abroad through strategic alliances. Although these results seem to support the information asymmetry perspective, they can also be explained taking into account the fact that telecom firms usually have the strongest market position in their home market. Due to this unique position, through domestic acquisitions, these firms can gain rents that are not available to most of the remaining bidders. Acquisitions in this industry seem to be a way for firms with high q values to take advantage of the market position in their home country.

This work has some limitations. First, it would be very interesting to have more detailed information about the target firms involved in acquisitions, and also about all partners that sign each strategic alliance. This information could shed more light in terms of the net value creation of each kind of business combination. However, as some of those firms were not listed firms, we focused our attention on the EU telecom companies so as not to reduce our sample size. Nonetheless, this did not permit us to introduce some interesting controls, such as the relative size of the companies involved. Another limitation is that the effect of management costs and the cultural distance deserves more detailed research. In fact, more is needed, although it was not available.

Finally, much more work remains to be done to identify the performance consequences of alliances and acquisitions aimed at expanding firm boundaries. Although we have controlled for unobserved heterogeneity, we did not find a clear pattern of reaction in the stock markets. More detailed studies combining event study techniques with qualitative data should be helpful to advance in this area (Trillas, 2002).

Acknowledgments

Financial support from the European Union (Fondo Europeo de Desarrollo Regional—FEDR) and the Spanish Ministerio de Ciencia y Tecnología (Project SCE 2003-08069) are gratefully acknowledged.

References

Balakrishnan, S., & Koza, M. (1993). Information asymmetry, adverse selection and joint ventures. *Journal of Economic Behavior and Organization, 20,* 99-117.

Bradley, M., Desai, A., & Kim, E. (1988). Synergistic gains from corporate acquisitions and their division between the stockholders of target and acquiring firms. *Journal of Financial Economics, 21*(1), 3-40.

Brooks, L. D., Feils, D. J., & Sahoo, B. K. (2000). The impact of shifts in forecasted earnings and systematic risk on acquiring firm shareholder wealth in domestic and international acquisitions. *Journal of Financial and Strategic Decisions, 13*(3), 1-7.

Brown, S. J., & Warner, J. (1985). Using daily stock returns. *Journal of Financial Economics, 14,* 3-31.

Cakici, N., Hessel, C., & Tandon, K. (1996). Foreign acquisitions in the United States: Effect on shareholder wealth of foreign acquiring firms. *Journal of Banking and Finance, 20,* 307-329.

Campa, J. M., & Kedia, S. (2002). Explaining the diversification discount. *Journal of Finance, 57*(4), 1731-1762.

Chan, S. H., Kensinger, A. J., Keown, A., & Martin, J. D. (1997). Do strategic alliances create value? *Journal of Financial Economics, 46,* 199-221.

Chatterjee, S. (1986). Types of synergy and economic value: The impact of acquisitions on merging and rival firms. *Strategic Management Journal, 7,* 119-139.

Chatterjee, S., Lubatkin, M. H., Schweiger, D. M., & Weber, Y. (1992). Cultural differences and shareholder value in related mergers: Linking equity and human capital. *Strategic Management Journal, 13,* 319-334.

Chen, H., Hu, M. Y., & Shieh, J. (1991). The wealth effect of international joint ventures: The case of U.S. investment in China. *Financial Management,4,* 31-41.

Chen, S., Ho, K., Lee, C., & Yeo, G. (2000). Investment opportunities, free cash flow and market reaction to international joint ventures. *Journal of Banking and Finance, 24,* 1747-1765.

Chung, I., Koford, K., & Lee, I. (1993). Stock market views of corporate multinationalism: Some evidence from announcements of international joint ventures. *The Quarterly Review of Economics and Finance, 35*(5), 275-293.

Chung, K. H., & Pruitt, S. W. (1994). A simple approximation of Tobin's Q. *Financial Management, 20*(7), 70-74.

Crutchley, C., Guo, E., & Hansen, R. (1991). Stockholder benefits from Japanese-U.S. joint ventures. *Financial Management, 4*, 22-30.

Das, S., Sen, P., & Sengupta, S. (1998). Impact of strategic alliances on firm valuation. *Academy of Management Journal, 41*(1), 27-41.

Datta, D. K., & Puia, G. (1995). Cross-border acquisitions: An explanation of the influence of relatedness and cultural fit on shareholder value creation in U.S. acquiring firms. *Management International Review, 35*(4), 337-359.

Delios, A., & Beamish, P. W. (1999). Geographic scope, product diversification, and the corporate performance of Japanese firms. *Strategic Management Journal, 20,* 711-727.

DeLong, G. L. (2001). Stockholder gains from focusing versus diversifying bank mergers. *Journal of Financial Economics, 59,* 221-252.

Dewenter, K. L. (1995). Does market react differently to domestic and foreign takeover announcements? Evidence from the U.S. chemical and retail industries. *Journal of Financial Economics, 37,* 421-441.

Doukas, J., & Travlos, N. G. (1988). The effect of corporate multinationalism on shareholders' wealth: Evidence from international acquisitions. *The Journal of Finance, 43*(5), 1161-1175.

Eckbo, B. E., & Thorburn, K. S. (2000). Gains to bidder firms revisited: Domestic and foreign acquisitions in Canada. *Journal of Financial and Quantitative Analysis, 35*(1), 1-25.

Finnerty, J., Owers, J., & Rogers, R. (1986). The valuation impact of joint ventures. *Management International Review, 26*(2), 14-26.

Flanagan, D. J. (1996). Announcements of purely related and purely unrelated mergers and shareholder returns: Reconciling the relatedness paradox. *Journal of Management, 22*(6), 823-835.

Frame, W. S., & Lastrapes, W. D. (1998). Abnormal returns in the acquisition market: The case of bank holding companies, 1990-1993. *Journal of Financial Services Research, 14*(2), 145-163.

Fröhls, M., Keown, A., McNabb, M., & Martin, J. (1998). Growth opportunities, corporate governance and the market value of multinational joint ventures. *Managerial and Decision Economics, 19,* 13-29.

García-Canal, E. (1996). Contractual form in domestic and international strategic alliances. *Organization Studies, 17,* 773-794.

García-Canal, E., Sánchez-Lorda, P., & Vidal, M. (2002). Reacción del mercado de capitales ante los procesos de combinación empresarial: Un panorama. (Stock market reaction to business combinations: A survey.) *Revista Europea de Dirección y Economía de la Empresa, 11*(4), 111-134.

García-Canal, E., Valdés-Llaneza, A., & Ariño, A. (2003). Effectiveness of dyadic and multi-party joint ventures. *Organization Studies, 24*(5), 743-770.

Gulati, R. (1995). Does familiarity breed trust? The implications of repeated ties for contractual choice in alliances. *Academy of Management Journal, 38*(1), 85-112.

Hall, R. E. (2001) The stock market and capital accumulation. *American Economic Review, 91,* 1185-1202.

Hamel, G. (1991) Competition for competence and interpartner learning within international strategic alliances. *Strategic Management Journal, 12,* 83-103.

Harrison, J. S., Hitt, M. A., Hoskisson, R. E., & Ireland, R. D. (1991). Synergies and post-acquisition performance: Similarities versus differences in resource allocations. *Journal of Management, 17,* 173-190.

Harrison, J. S., Hitt, M. A., Hoskisson, R. E., & Ireland, R. D. (2001). Resource complementarity in business combinations: Extending the logic to organizational alliances. *Journal of Management, 27,* 679-690.

Hennart, J. F., & Park, Y. (1993). Greenfield vs. acquisition: The strategy of Japanese investors in the United States. *Management Science, 39,* 1054-1070.

Hennart, J. F., & Reddy, S. (1997). The choice between mergers/acquisitions and joint ventures: The case of Japanese investors in the United States. *Strategic Management Journal, 18,* 1-12.

Hofstede, G. (1980). *Culture's consequences: International differences on work-related values.* Beverly Hills: Sage Publications.

Hofstede, G. (2001). *Culture's consequences: International differences in work-related values.* Beverly Hills: Sage Publications.

Hubbard, R. G., & Palia, D. (1999). A reexamination of the conglomerate merger wave in the 1960s: An internal capital markets view. *The Journal of Finance, 54*(3), 1131-1152.

Kim, K., & Park, J. (2002). The determinants of value creation for partner firms in the global alliance context. *Management International Review, 42*(4), 361-384.

Kogut, B. (1991). Joint ventures and the option to expand and acquire. *Management Science, 37,* 19-33.

Kogut, B., & Singh, H. (1988). The effect of national culture on the choice of entry mode. *Journal of International Business Studies, 19,* 411-432.

Koh, J., & Venkatraman, N. (1991). Joint venture formations and stock market reactions: An assessment in the information technology sector. *Academy of Management Journal, 34*(4), 869-892.

Kuipers, D. R., Miller, D., & Patel, A. (2003). *The legal environment and corporate valuation: Evidence from cross-border takeovers.* Working Paper. Lubbock, Texas: Texas Tech University.

Leeth, J. D., & Borg, R. (2000). The impact of takeovers on shareholder wealth during the 1920s merger wave. *Journal of Financial and Quantitative Analysis, 35*(2), 217-238.

López-Duarte, C., & García-Canal, E. (2002). Adverse selection and the choice between joint ventures and acquisitions: Evidence from Spanish firms. *Journal of Institutional and Theoretical Economics, 158,* 304-324.

Low, A. L. (2001). *A study of two-step spinoffs.* Working Paper. New York: The Leonard N. Stern School of Business.

Lubatkin, M. H. (1987). Merger strategies and stockholder value. *Strategic Management Journal, 8,* 39-53.

Lummer, S., & McConnell, J. J. (1990). Stock valuation effects on international joint ventures. In J. Rhee & L. Chand (Eds.), *Pacific Basin Capital Markets Research* (pp. 531-546). New York: Elsevier Science Publishers.

McWilliams, A., & Siegel, D. (1997). Event studies in management research: Theoretical and empirical issues. *Academy of Management Journal, 40*(3), 626-657.

Merchant, H., & Schendel, D. (2000). How do international joint ventures create shareholder value? *Strategic Management Journal, 21,* 723-737.

Mohanram, P., & Nanda, A. (1996). When do joint ventures create value. *Academy of Management Proceedings.* AL: Academy of Management.

Morck, R., Shleifer, A., & Vishny, R. W. (1990). Do managerial objectives drive bad acquisitions? *The Journal of Finance, 44*(1), 31-48.

Morck, R., & Yeung, B. (1992). Internalization: An event study test. *Journal of International Economics, 33,* 41-56.

Ojah, K., Seitz, N., & Rawashdeh, M. (1997). Why manufacture offshore? An empirical analysis of valuation effects. *The Financial Review, 32,* 477-499.

Oxley, J. E. (1997). Appropriability hazards and governance in strategic alliances: A transaction cost approach. *Journal of Law, Economics and Organization, 13,* 387-409.

Pantzalis, C., & Kohers, N. (2000). *The effect of legal environment and foreign operations network on U. S. acquirer gains in foreign takeovers.* Working Paper. Tampa, FL: University of South Florida.

Park, S., & Kim, D. (1997). Market valuation of joint ventures: Joint venture characteristics and wealth gains. *Journal of Business Venturing, 12,* 83-108.

Parkhe, A. (1993). Partner nationality and the structure-performance relationship in strategic alliances. *Organization Science, 4,* 301-324.

Rad, A. T., & van Beek, L. (1999). Market valuation of European bank mergers. *European Management Journal, 17*(5), 532-540.

Reuer, J., & Koza, M. (2000). Asymmetric information and joint venture performance: Theory and evidence for domestic and international joint ventures. *Strategic Management Journal, 21,* 81-88.

Reuer, J., Park, K. M., & Zollo, M. (2002). Experiential learning in international joint ventures: The roles of experience heterogeneity and venture novelty. In F. Contractor & P. Lorange (Eds.), *Cooperative strategies and alliances* (pp. 321-346). Oxford: Elsevier Science.

Seth, A. (1990). Sources of value creation in acquisitions: An empirical investigation. *Strategic Management Journal, 11,* 431-446.

Sharpe, S. (1964). Capital asset pricing: A theory of market equilibrium under conditions of risk. *Journal of Finance, 19,* 425-442.

Shelton, L. M. (2000). Merger market dynamics: Insights into the behavior of target and bidder firms. *Journal of Economic Behavior and Organization, 41*(4), 363-383.

Singh, H., & Montgomery, C. (1987). Corporate acquisition strategies and economic performance. *Strategic Management Journal, 8,* 377-386.

Trillas, F. (2002). Mergers, acquisitions and the control of telecommunications firms in Europe. *Telecommunications Policy, 26,* 169-186.

Very, P., Lubatkin, M., Calori, R., & Veiga, J. (1997). Relative standing and the performance of recently acquired European firms. *Strategic Management Journal, 18,* 593-614.

Vidal, M. (2000). Las alianzas estratégicas globales para la internacionalización. Su contribución a la creación de valor en la empresa. (Global strategic alliances to internationalize: Its contribution to value creation in the firm.)*Economía Industrial, 333,* 49-56.

Vidal, M., & García-Canal, E. (2003). Complementarity and leverage as drivers of the stock market reaction to global alliance formation. *Long Range Planning, 36*(6), 565-578.

Villalonga, B. (2004). Does diversification cause the diversification discount? *Financial Management, 33,* 5-27.

Waheed, A., & Mathur, I. (1995). Wealth effects of foreign expansion by U.S. banks. *Journal of Banking and Finance, 19,* 823-842.

Wu, C., & Wei, K. C. (1998). Cooperative R&D and the value of the firm. *Review of Industrial Organization, 13,* 425-446.

Endnotes

1 Abnormal returns are the variation in the stock price of a company directly attributable to a specific event, in this case, the formation of a business combination.

2 Cakici, Hessel, and Tandon, (1996); Eckbo and Thorburn (2000); or Pantzalis and Kohers (2000), among others, find these results.

3 See, for instance, Koh and Venkatraman (1991).

4 Chen, Ho, Lee, and Yeo, (2000); Fröhls, Keown, McNabb, and Martin, (1998).

5 This section and the next one are based, in part, in García-Canal et al. (2002).

6 We also included in the sample those firms classified under the SIC 4899 (communications services, not elsewhere classified) and described in the database under the field *Acquiror Short Business Description* as providers of telecommunications services. That was the case of British telecom and Mannesmann.

7 This database contains, among many other data, the stock prices of all the companies publicly traded in all the major European stock exchanges. For our study, the relevant data of this database is the Total Return Index, which reflects the variation in the stock price of each company in relation to a specific date. This index is already prepared to conduct event studies and is corrected for dividend payments and equity operations. Therefore, this information is very useful in order to develop an event study.

8 Low (2001) and Reuer, Park, and Zollo, (2002), among others, have shown that the date that figures in SDC does not always coincide with the first day in which the announcement of a business combination becomes public. Since the exact measure of this date is crucial for a correct application of the methodology described in this chapter, we verified all the announcement dates by means of systematic manual searches in that database.

9 Following McWilliams and Siegel (1997), we eliminated from the original sample all those observations in which there were further news concerning seasoned equity offerings, stock reductions, dividend payments, contracts with the State, other acquisitions or alliances different from those considered, or decisions concerning changes in a key executive of the studied companies.

10 We used as reference index the *Global Index* provided by DataStream. This is a market index for each stock market and is built using the same criteria for each one of them. Thus, the index includes the main listed firms in each market in terms of market capitalization (120 titles on average in each case) and is corrected by dividend payments as well as by seasoned equity offerings or stock reductions.

11 Tobin's Q has been widely used as a measure of the value of the intangible assets of the firms (Morck, Shleifer, & Vishny, 1990; Morck & Yeung, 1992). However, according to Hall (2001), it also can be reflecting the market power of the company. In the case of the telecommunications industry, in which there are important regulatory entry barriers, Tobin's Q can be understood as a measure of the strength of the market position of a firm in its current market.

12 See, for instance, Vidal and García-Canal (2003).

13 This argument is in line with the literature that studies alliances as real options (see Kogut, 1991).

Chapter XI

Strategic Alliances for Environmental Adaptation: The Support of Virtual Networks

César Camisón, Universitat Jaume I, Spain

Abstract

Although the literature on strategic alliances defines the role of information technologies (IT) as important, any reasoned explanation of it is still anecdotic. This study focuses on a particular type of technological relationship between firms and institutions for developing environmental management innovations through virtual cooperation. It goes into particular depth on the economic and strategic reasons explaining the formation of alliances based on IT. The development of environmental adaptation strategies can be effectively supported with public initiatives for providing advanced, knowledge-intensive services through the creation of technological extension infrastructures. The market's failure to provide useful public goods for improving environmental development may be corrected effectively using virtual inter-organizational networks facilitating collective learning. This study presents an example of a regional initiative for a cooperative platform based on IT that can result in the assimilation of technological and organizational competences in environmental management by firms located in the area.

Introduction

A common line of thought in current strategy literature is that the innovations introduced by companies are largely based either on collective processes in which firms, institutions, and other agents located in a given geographical context take part or on inter-organizational cooperation agreements. Strategic alliances with institutions, competitors, consumers and suppliers, whether or not they are local, appear to be increasingly important for strategic reasons (they generate value by producing sustainable competitive advantages) and for financial ones (they reduce transaction and production costs). By contrast, despite the fact that the literature on networks defines the role of IT as important, the explanation of the causes of this is still anecdotic. This study focuses on a particular type of technological relationship between firms and institutions for developing environmental management innovations through virtual cooperation. This case study allows us to look in depth at the financial and strategic reasons explaining the formation and maintenance of networks based on IT.

Environmental adaptation falls within the social and strategic action of companies, and here the subject of social responsibility is probably the one considered by management to be most important (Owen & Scherer, 1993). But the adoption of environmentally committed attitudes is no longer merely a matter of ethics and social responsibility. Instead, it has grown to become a spearhead for change towards business management based on the concept of sustainability (Andersson & Wolff, 1996; Jennings & Zandbergen, 1995; Roome, 1992; Shrivastava, 1995a, 1995b, 1995c). Companies that do not internalize environmental considerations will have little chance in the future, when the changes have to be made more urgently (Vastag, Kerekes, & Rondinelli, 1996).

However, there is a large group of firms that have problems adapting in an environmentally proactive way. The development of environmental adaptation strategies may be effectively supported with public initiatives for providing advanced, knowledge-intensive services through the creation of technology extension services. In particular, public policies based on inter-organizational networks have been studied as effective actions for correcting market failures to provide useful public goods for improving the environmental performance of firms located in a particular area. The collective learning that takes place between companies that are geographically close to one other, with institutional support, can be intensified with the integration of the network on IT. The Internet is an extremely important technology (Porter, 2001, p. 63), which is transforming the way in which firms and institutions interact. This technological change has put within the reach of organizations a new and very interesting network structure for incentivizing proactive strategies for improving and protecting the natural environment, running parallel with regulation and instruments based on the market but without their asymmetrical effects.

This study establishes a relationship between the creation of strategic virtual networks and the effectiveness of environmental adaptation. The presentation is based on a regional initiative made up of a cooperative platform based on IT that can result in the companies making up the network and assimilating technological and organizational competences in environmental management. It is Ecosigma, a non-profit association founded in October 2003 by five important agents from Valencian Community. Its mission is to develop, publicize, and maintain the Ecosigma system as an innovative environmental Web site providing virtual support for the application of tools and principles respectful towards the natural environment. This project offers a valuable opportunity for observing how knowledge and information (in this case for environmental adaptation) are shared and disseminated within a virtual cooperation platform model. This strategic alliance based on IT has provided effective solutions to problems companies suffer in environmental adaptation.

The first part of this chapter consists of a discussion on the theorization of the problem of environmental adaptation and virtual cooperation. We will then deal with environmental adaptation strategies and possible cooperative or collective learning solutions as an effective way of achieving it. Our attention is concentrated on the value of strategic virtual networks based on IT. Finally, we prepare the design for a regional initiative for developing is cooperative platform based on IT, which would remove the barriers preventing companies undertaking proactive environmental strategies.

The Problem of Environmental Adaptation and Cooperation as Proactive Environmental Strategy

There are various reasons for the inclusion of the environmental dimension on the agenda of fundamental issues for management (Bansal & Roth, 2000). The first explanation stems from the generalization of concern for the natural environment. In many industries it is interpreted as an entry barrier, a requirement for *green* market segments, or a legitimacy requirement by many stakeholders. It is also the result of increasingly strict public regulation (including regulatory pressure and the risk of penalties for legal breaches) resulting from growing awareness and social demand and from an environment with serious problems requiring immediate action.

The very profitability and competitive advantages potentially associated with environmental investments make up another solid argument in their favour. The literature emphasizes the fact that environmental conditions affect a company's competitiveness because it is a source not only of threats but also of opportunities (Sharma, 2000). The value of a proactive environmental strategy would also stem from the development of organizational competences, which are a source of lasting competitive advantages (Aragón-Correa, 1998; Aragón-Correa & Sharma, 2003; Christmann, 2000; Porter & Van der Linde, 1995; Sharma & Vrendenberg, 1998; Shrivastava, 1995b). Hart (1995) has identified in environmental strategies (specifically of three: pollution prevention, product monitoring, and clean technologies) the characteristics that turn them into sustainable competitive advantages: their tacit nature, social complexity, and specific nature. The solutions allowing environmental impact to be reduced and competitive advantages to be simultaneously created are called "win-win" solutions (Florida, 1996, p. 81) and even "win-win-win" ones, indicating that the company, the environment, and consumers all win (Elkington, 1994, p. 90).

A broad thrust of research (Azzone, Bértele, & Noci, 1997, p. 562; Schot, Brand, & Fischer, 1997, p. 155) maintains that environmental adaptation may contribute towards the achievement of both cost advantages and differentiation. The most often attempted methods of developing differentiation based on environmental adaptation are the use of legal provisions compelling the adoption of certain technologies to create a turnkey market for patentable innovations, and the seeking of *green* markets. Meanwhile, cost advantages can come from the efficiency gain achievable with the introduction of clean technology innovations (Haustein, 1985); savings on operating costs (Porter & Van der Linde, 1995, p. 125)—both these measures included under the heading of eco-efficiency—the reduction in risks associated with responsibility, and the reduction in the cost of environmental impacts.

Another approach, when it comes to studying the factors determining firms' patterns of environmental adaptation, is management perception of the environmental attributes of priority interest groups. The literature (i.e., Andersson & Wolff, 1996; Post, 1994) has identified two stakeholders whose environmental demands are a priority for management: public administrations and consumers. In this study, we are going to focus on the effects of public environmental policies. The effect of administrations on the environmental behaviour of economic agents is clear in regulatory actions such as legislative development;

exhaustive and demanding environmental surveillance; and control policies. The economic instruments public bodies can use to stimulate environmental adaptation by firms are many and well-known (Van der Linde, 1993).

However, public action is not enough for environmental adaptation to penetrate the behaviour of private agents successfully. Regulation and market-based strategies to protect the natural environment may not be sufficient to induce sustainable economic development. The activation of a great deal of knowledge about clean technologies and best environmental management practices is now required, and this cannot be entirely supplied by one agent working alone. The adoption of proactive environmental adaptation is seen to be delayed by the inefficiency of the environmental information market. The failures in this market can be seen in the technical and legislative difficulties, which are closely linked to problems of information asymmetry. The information and knowledge available about the problem of environmental adaptation suffer from a high level of uncertainty (Cothern & Ross, 1994, p. 2), stemming from the very complexity of environmental considerations (Roome, 1992, p. 11-13); the limitations of scientific knowledge (Rejeski, 1994, p. 270); the fact that adequate solutions to some of these problems do not exist (Shrivastava, 1995b, p. 196); and the deficiencies of internal information and monitoring systems (Franz & Hefter, 1999). Dealing with these problems is a task requiring a combination of technical and management skills (Roome, 1994, p. 67). Authors like Shrivastava (1995a, p. 134; 1995c, p. 957) come to postulate that in order to adapt themselves to the needs of sustainable development, companies must rethink their objectives, strategies, structures, concepts of performance, and interpretations of the environment. The complexity of environmental development also extends to the simple comprehension of the actions, systems, and means available, which suffers from the so-called "Tower of Babel Syndrome" (SustainAbility, SPOLD, & BiE, 1993, p. 12).

The problem of information monitoring is crucial for environmental management: information for the creation or discovery of new ecologically created niches; information for understanding and responding to new consumer aspirations in the environmental sphere; information for discovering best practices in environmental management among the technologies and management systems existing on the market; information for adapting to new market requirements concerning environmental certification, management, or regulation systems. We are talking about technological challenges (new products and processes), market challenges (changes in structure and nature), and sociological challenges (change in the value systems of consumers, citizens, institutions, and members of the organization). In summary, information is needed in order to achieve greater flexibility and innovation. However, it must not be forgotten that too much information can be as much of a problem as too little. Rivero (1991) has emphasized the need to ensure access to really important information and to try to get it in a "pre-processed" form. The definition of monitoring and the selection of information as a basic nutrient for competitiveness means that in order to survive and grow, firms must develop new systems for capturing information and new skills for filtering, evaluating, and using the information coming from their environment.

Faced with the impossibility of generating all the appropriate information internally, firms are often driven to obtain it on the market. Business services are a plural set of activities whose aim is the external provision of value chain activities with a view to improving effectiveness. Environmental adaptation services are a category of advanced, knowledge-intensive business services. Here we find another inefficiency in the information market: the difficulties the market alone finds in generating the required range of advanced or knowledge-intensive services. This general problem shows itself particularly in the fact that a great deal of environmental information cannot be on the market. Following Brusco (1996), we can find various reasons for this market failure. Firstly, the technical knowledge required for providing real environmental adaptation services can often not be found in the social environment in which the firm is operating and therefore has to be brought in from outside. Secondly, although the organization social environment may offer this specialized knowledge, the production of real services of this kind demands the availability of qualified staff together with a great investment, which often takes considerable time to bring in a real return. Thirdly, the very nature of the information market often makes it difficult for companies to sell information with commercial criteria. The external component accompanying the provision of knowledge-intensive services indicates that its impact on the environment exceeds the gains that can be obtained by the firm making the investment in order to enjoy them. The fact that a good part of the information required is a public good means it can be enjoyed free by other organizations, which discourages firms from individually committing resources to producing it.

The need for information about environmental technologies, markets, and regulation, which is difficult to obtain either internally or externally, incentivizes the development of new forms of cooperation outside the market and the hierarchy, which may contribute to creating these collective goods. Strategic alliances have been widely dealt with in the literature, although largely referring to productive and commercial activities. By contrast, they have scarcely been covered in the literature on environmental management, despite the fact that the establishment of cooperation relationships with other economic agents has been one of the most popular routes for environmental adaptation, either in the form of inter-business alliances (Biddle, 1993, p. 145; Hartman & Stafford, 1997; Schmidheiny, 1992, p. 145-148; Welford, 1995, p. 40) or in the form of collaboration with non-profit organizations (Hartman & Stafford, 1997; Prince & Denison, 1994).

There are a wide variety of networks. In this study we are especially interested in the type with the greatest potential for value creation, strategic networks. These are agreements under which several organizations interact through a web of stable relationships allowing products or services to be provided in a coordinated way, offering many of the advantages of a vertically integrated company and of resorting to the market, while avoiding some of the disadvantages (Jarillo, 1993). The value created using strategic networks depends on their use as a mechanism for obtaining competences (particularly knowledge-based ones), which improve the partners' competitiveness (Dyer & Singh, 1998; Hamel, 1991; Nooteboom, De Jong, Vossen, Helper, & Sako, 2000; Oliver, 1997). The strategic logic of cooperation also lies in its being an organizational learning process (through the factual internalization of another company's competences) to improve the firm's own competitive advantages (Pucik, 1988); as a means for accessing strategic assets such as specialized management knowledge (Castanias & Helfat, 1991); as an instrument for flexible adaptation to turbulent environments (Young & Wilkinson, 1997), and as useful cement for introducing organizational changes (Nooteboom et al., 2000), making internal barriers permeable to innovation (Roy, Sivakumar, & Wilkinson, 2004).

Physical location can have a significant impact on the generation of innovation, as the knowledge tends to be "sticky" (Ogawa, 1998). Given the complex and interactive (co-production of advanced services) nature of innovation (Lundvall & Johnson, 1994) and that learning is a localized process, which is the result of specific territorial histories (Asheim & Isaksen, 2000; Isaksen & Hauge, 2002), capacity for business innovation may be encouraged by the proximity between the components of the system. In the last decades of the 20th century, the literature on regional development emphasized the crucial role of local and regional resources for territorially based economic and social development (Maillat, 1998; Thierstein & Walser, 1997). Consequently, local and regional levels have been identified as important starting points for development strategies and environmental policy, as experiments like Agenda 21 have shown (Ekins & Newby, 1998; Gibbs, 1996).

However, it is generally admitted that a development model of this kind needs the stimulus of public strategies (Seidl, Schelske, Joshi, & Jenny, 2003, p. 335). Public intervention is seen to be justified by market failures to provide environmental adaptation services. The institutional context must therefore be competent in promoting these collective goods. The autonomous regional administration alone cannot supply all the institutions capable of meeting companies' environmental adaptation needs. But the public administration can provide the territorial organization with inter-organizational networks—so much the better if they are extra-territorial and international—including industrial companies, organizational services, consultancy firms, and local and regional institutions (Muller, 2001). Public action must stimulate network organization that can bring the local and regional resources available in the area together with the technological and organizational environmental management capabilities of institutions specializing in each task. These networks must offer more facilities for monitoring and exchanging information and knowledge in order to encourage environmental adaptation; innovation in products, processes, and organizational systems, with respect to the natural environment; the adoption of new clean technologies; and a human capital with the right attitude and qualifications, in this respect. Public intervention must then serve to incentivize the creation of an organization/innovation/knowledge circle, by means of which both coded and tacit knowledge concerning environmental adaptation can be assimilated by local productive structures, being incorporated into their contextual knowledge.

Collective Learning in Strategic Alliances Based on IT

The role of IT as a creator of value and of competitive advantages has been widely recognized in the literature (Barney, 1991; Cash & Konsynski, 1985; Clemons, 1991, 1986; Clemons & Row, 1991; Feeny & Ives, 1990; Mata, Fuerst, & Barney, 1995; McFarlan, 1984). The financial and strategic consequences of IT on the hierarchy and the market have been widely dissected. By contrast, the literature on networks shows an important gap. Although IT is defined as theoretically important in training and in value creation and there are even predictions that its value leverage will be greater in this organizational form, empirical evidence of the reasons for this is still insufficient (i.e., Camisón & Lapiedra, 1999a; Bensaou, 1994).

The positive impact of IT on the effectiveness of strategic networks stems from both financial and strategic reasons. To analyse them, we must differentiate two dimensions: (1) the dimension of IT as a resource, referring to IT infrastructures (ITI) (tangible assets) and (2) its dimension as a capability, referring to organizational routines and management skills in dealing with these technologies (ITMAN). The competitive potential of both dimensions is very different. In time, all competitors in an industry are capable of implementing the same ITI, so the generation of sustainable competitive advantages lies in developing intangible information and knowledge management capabilities (Ross, Beath, & Goodhue, 1996). Strategic assets are created through acquiring experience in using IT and not through merely possessing them. This can be the source of the difficulties companies can have in using IT competitively (Ba, Whinston, & Zhang, 1999).

The positive financial impact of IT on the formation and maintenance of networks is based on their reducing network structure transaction costs.[1]

Following Clemons and Row's (1992) approach to the division of transaction costs, we maintain that the implementation of IT (both ITI and ITMAN) is translated into value creation for the networks because it contributes towards reducing the costs of coordinating the relationship (streamlining its information channel components and management process) and the transaction risks implicit in potential opportunist behaviour.[2]

Although network and hierarchical structures are different forms of organization, they share certain aspects. Among their points in common are the organizational coordination mechanisms created for dealing with the growing load of information because of uncertainty over the tasks to be coordinated. The existence of common coordination patterns stems from the notion of coordination as the series of protocols, tasks, and decision-making mechanisms designed for carrying out concerted actions between interdependent units (Thompson, 1967). The coordination of the interdependent units making up the hierarchy and the coordination of the agreed activities in a network between interdependent but independent organizations can be undertaken using the same coordination mechanisms, with slight alterations. Following Mintzberg's (1979) classic typology, the coordination of inter-organizational networks can be supported by five mechanisms: (1) mutual adaptation between the firms making up the network; (2) direct supervision of the agreed activities; (3) the standardization of firms' skills so they can be integrated into the network; (4) the standardization of the network firms' working processes; and (5) the standardization of results.

IT can contribute towards reducing coordination costs by providing support for putting each of these mechanisms into practice, improving the lateral links that facilitate each option. IT contributes toward reducing network coordination costs when the network is coordinated by mutual adaptation because of their support for improving the lateral links that facilitate it. The lateral links facilitating mutual adaptation, which benefit from the implementation of IT, are decision making and communication processes crossing hierarchical lines and aiding direct contact between people and working in groups (Clemons & Row, 1992; Kumar & van Dissel, 1996; Rockart & Short, 1989). In addition, IT can contribute to reducing coordination costs when the network is organized based on the two least structured mechanisms—mutual adaptation and direct supervision—as they streamline the development of organizational routines for the structured problems of information management and management of shared activities, lightening the coordination workload, which is reduced to the less structured problems. IT also contributes towards network coordination costs when it is coordinated through the standardization of results by improving the effectiveness of the lateral links that facilitate it, such as the planning and monitoring of shared activities. Planning benefits from the power of IT to compile; process (making complex calculations, simulations and models) and distribute data; and for communication. Monitoring benefits from the facilities IT provide; for supervising geographically distant units; detecting deviations and their causes (at any level of aggregation); drawing up complex indicators; and drawing up both normal and exceptional reports. IT can also contribute towards reducing coordination costs when the network is organized on the basis of the standardization of processes and skills. The involvement of IT can support the increased efficiency of related processes, as would be the case with electronic data interchange (EDI) or shared skills, as would happen with virtual learning models or the management of knowledge shared on the network (Camisón & Lapiedra, 199b).

The efficiency of the network could be reduced, if this form of organization leads to a growth of the probability of opportunist behaviour by a member. The reduction in coordination costs could then be cancelled out by the increase in costs associated with the increased risk of the transaction. However, the second financial advantage of IT is that they allow transaction risks in inter-organizational networks to be reduced. The literature identifies three possible sources of transaction risk (Clemons and Row, 1992): (1) specialized investments with little value outside the network (Klein, Crawford, & Alchian, 1978; Williamson, 1975); (2) information asymmetries making the monitoring of the performance of network members difficult; and (3) the leakage of assets (basically information and knowledge), which were made available to the network; for example when a member of the network learns more than initially agreed from another member of the network, which can result in use against the owner firm (Tsang, 2000; Williamson, 1991). IT has effects on specialized investments, the information asymmetries and the leakage of shared assets. As for specialized investments, it must be noted that there is a progressive tendency towards the use of standard ITI which does not involve the organization incurring sunk costs if it wants to leave the network and allows it to take advantage of the externalities produced by these activities. The hypothesis that standard ITI reduce transaction risk from specialized investments can therefore be suggested, together with the converse, that non-standard ITI increase it. A corollary of this premise is that firms may avoid integration into networks that make specialized investments in ITI compulsory, given that they increase transaction risk. Concerning information asymmetries, it can be hypothesized that IT (both in their tangible and intangible dimensions) reduce them, by clearly monitoring the performance of all network members (Clemons & Row, 1992; Kumar & van Dissel, 1996; Rockart & Short, 1989). By contrast, the third problem, the increase of transaction risk resulting from the leakage of assets made available to the network (particularly information and knowledge), can be increased with the use of IT, as these multiply access facilities (Clemons & Row, 1992). Consequently, the net effect of IT on transaction risks implicit in potential opportunist behaviour will depend on the balance between their effect on the three risk sources analysed. In addition, it must be borne in mind that the possible increase in transaction risk can be ameliorated or compensated by the continuity of transactions within stable inter-organizational networks, in which long-term relationships are established (Gulati, Nohria, & Zaheer, 2000; Madhok, 1995; Tsang, 2000) and where trust acquires central importance (Gulati et al., 2000; Jarillo, 1988; Tsang, 2000).

In addition, the quality of IT levering the value generated by networks stems from their strengthening of the effectiveness of network behaviour with the development of unique, inimitable assets. The singularity and inimitability of IT is based on the complementary and specific nature of the resources and capabilities attached to network relationships, defining them as strategic assets and, consequently, as a source of sustainable competitive advantages.

The formation and maintenance of networks is facilitated because IT (both ITI and ITMAN) acts as complementary resources and capacities supporting value creation strategies driven by the cooperation agreement. The resources and capabilities linked to IT are complementary assets, which provides them with the first condition for being classed as strategic assets (Milgrom & Roberts, 1995). In order to obtain, exploit, and develop competences using networks as opposed to by individual internal effort, a series of difficulties must be overcome: the problems of managing a set of firms as if they were a single organization; the polychromatic nature of activities shared by the network; the problems of information management; the need for compatibility between firms' organizational and cultural systems; and the geographical dispersal of members. IT can provide solutions to several of these obstacles. Network management can be improved by IT, as ITI and GESTIC are complementary assets for the planning, carrying out, and monitoring of network activities. This hypothesis does no more than extrapolate to inter-organizational relations the demonstrated support that IT provides for planning, carrying out, and monitoring activities by individual firms (Porter & Millar, 1985). Concerning the second problem, the more complex and diverse the activities shared on the network, the stronger the complementary nature of IT will be. The importance of the complementary nature of IT will equally be directly related to the—intangible—information component of the shared activities. The coordination and communication problems arising from geographical dispersion, which may be an obstacle to the development of the network, are also attenuated with IT. One final aspect that directly conditions the competitive scope of the complementary nature of IT for the network is the compatibility between the ITI and GESTIC. Perhaps, the network problem where IT has least effect is that of reconciling organizational and cultural asymmetry. However, IT can also lead to reductions in production costs beyond the reach of the hierarchy and the market thanks to the synergy effect deriving from the complementary nature of resources and capabilities between organizations making up the network and association with firms with lower costs of this kind.

The inimitability of IT is also based on the fact that ITMAN are specific to networks. Networks are idiosyncratic forms of organization created using a path dependence process, which is difficult for competitors outside the network to imitate (Gulati et al., 2000). ITMAN reinforces the specific nature of the network as the routines and management skills of ITI are adapted to the particular features of the firms making it up and their form of integration. ITMAN barriers to imitation also stem from their germination over long periods of time, involving high levels of social complexity, and the fact that their operation shows a high level of causal ambiguity.

The differential adoption of IT by the various agents can result in the development of different technological platforms with different degrees of integration (Bowersox, Closs, & Stank, 2000; Cachon & Fisher, 2000; Kalakota & Robinson, 1999). The more completely the systems are integrated between agents, the more routine communications and transactions will be conducted online, freeing time and resources for knowledge creation (Dwyer, Schurr, & Oh, 1987; Roy et al., 2004). The learning theory (Brown & Duguid, 1991; Lave & Wenger, 1991) also suggests that in these circumstances most routine interactions are conducted in digital form and that personal interactions will involve new learning and knowledge development. However, the cooperative integration of IT may be a factor discouraging radical innovation (Chandy & Tellis, 2000; Kalakota & Whinston, 1999; Perry, Cavaye, & Coote, 2002) if the technological platform is rigid, in the face of unlearning costs and organizational inertia. By contrast, flexible virtual cooperation can provide interactions leading to radical innovations (Leifer, Colarelli O'Connor, & Rice, 2001). This is so much easier when the network includes practice communities from different professional and knowledge spheres (Braun, 2002).

However, the problems discussed in these networks are largely idiosyncratic and require a level of learning and knowledge only possessed by professional colleagues (Almeida & Kogut, 1997). A crucial problem for environmental adaptation is that a good part of specific knowledge is tacit and fragmented, lacking formal networks to supply it. This knowledge depends on many informal groups, professional organizations, and technical service firms, which in a fragmented way, capture tacit knowledge separately from explicit knowledge in this scientific field (Orr, 1990). Access to this specialized knowledge is difficult (Millar, Demaid & Quintas, 1997), as it frequently depends on ambiguous associations such as e-mail lists. The scientific and technical community that stores tacit knowledge may be inextricable and accessible only online. Because of this, these practice communities have flourished with the development of the Internet. This is the case, for example, of the interest groups that interact to solve problems without cost.

This makes it even more difficult for organizations to access specific and tacit knowledge in the environmental field. The public promotion of firm access to intensive knowledge services is fundamental for these purposes as the use of advanced services facilitates the connectivity and receptivity (Antonelli, 1998) of information and knowledge networks. Networks based on IT facilitate the assimilation of technological competences deposited in the network by content providers, especially professional services firms, which provide knowledge and technical services from various origins and locations. Through a cooperative computer platform, firms can access the technological competences of specialized companies in order to combine the general knowledge competences stored in these intensive organizations (with a more codifiable content) with their own tacit knowledge (which each firm can obtain based on its own specific experience). The knowledge required for the innovation process can then be obtained (Antonelli, 1999) in the environmental field.

Case Study: ECOSIGMA

The earliest forerunner of Ecosigma was a European project developed in 2002 by SEPIVA, a public company belonging to the Valencian regional government. The national non-profit association Ecosigma was set up in October 2003 in order to build, publicize, and maintain the Ecosigma system. The founder members were the regional government (represented by SEPIVA), the Valencian Regional Council of Official Chambers of Commerce, Industry and Shipping, and business organizations (represented by the Valencian Region Confederation of Business Organizations [COEPA], and the Federation of Spanish Footwear Industries [FICE]). The purpose of the association is to incorporate as members all business organizations and chambers of commerce in the region or in Spain as a whole, which could contribute to it achieving its mission. So, in 2004 the Higher Council of Spanish Chambers of Commerce, Industry and Shipping, and the Spanish business associations covering aggregates, natural stone, toys, footwear, timber and furniture, and ceramics (ASCER); metal (Confemetal); and monitoring bodies (Asorco) were included. Figure 1 shows the organizational structure of the project.

The purpose of this project is to develop a computer platform providing industrial firms in the Valencian Community (a Spanish region) with a tool for the environmental adaptation of industries and firms with problems in this area. The project has three environmental aims. Firstly, to provide enterprises with a self-diagnosis of their position as regards to current environmental regulations. A second aim was defined as providing knowledge to firms about how to adapt themselves to environmental regulations and how to make the most efficient possible use of the technology available on the market. These specific actions fall

Figure 1. Organization of the Ecosigma project

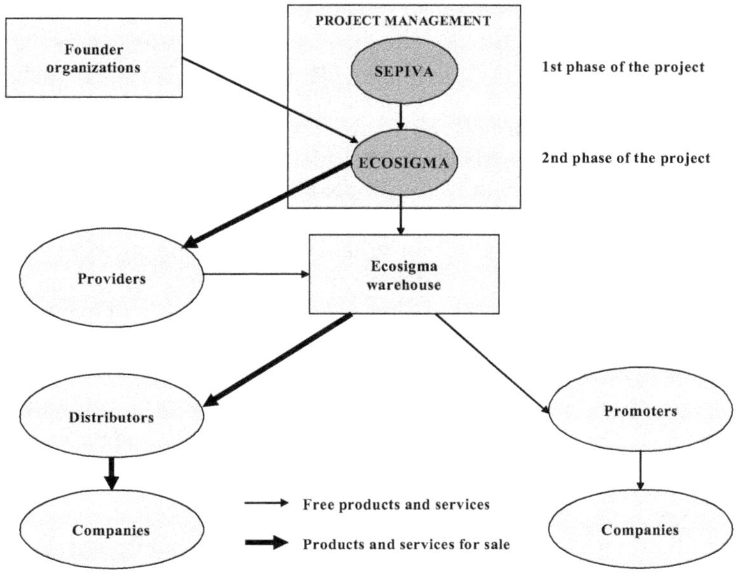

within a third objective, which is to optimize the environment of the firms and industrial districts within the Valencian Community, resulting in a cleaner, healthier natural environment. Financially, the project is seeking to recover as quickly as it can the possible extra costs involved in production under environmentally respectful regulations through the production and sale of products with high environmental quality. This financial objective must be based on access to the best environmental management practices in order to optimize processes, increase productivity, and achieve competitive advantages in *green* market segments.

To achieve these aims, the Ecosigma system has been set up as a landmark virtual network on environmental issues. The physical support for the network is the digital warehouse, where the catalogue of environmental services and tools is deposited. The Ecosigma system depends on a portal incorporating a web server implemented in ASP.NET and with a process logic implemented using .NET language in the form of DLLs for checking the rules of business and managing the MySQL-type database. The web application allows access from any computer connected to the Internet, both to the collected data and the system configuration.

The network far exceeds the scope of the public administration. The intention was that the services to be provided should be woven as deeply as possible into the regional industry, preventing duplication in the development of environmental content by business associations and other interface institutions, together with a willingness to involve all producer organizations and distributors of computer systems and applications for environmental adaptation. These reasons led to the design of a cooperation system where all the agents relevant to environmental problems were present: regional institutions, business associations, content providers (above all environmental consultancy service companies and information technology services firms), and companies as end users. So, it is a virtual cooperative platform for the assimilation and development by companies of the technological and organizational environmental management competences deposited in the network by the content providers.

The function of the network coordination body is to act as an interface between the components of the network, maximizing their interaction and the transfer of products (with or without image). There are four types of agents in the system:

- **Providers.** These are consultants specializing in the environment or in IT who develop the tools and services in the catalogue. These organizations can include their products in the catalogue after adapting them to the system so they can later be distributed to promoters, distributors and end users. There are different types of providers depending on the type of product developed. Regional or national federations of business associations can also be registered content providers.
- **Promoters.** These are organizations that acquire products and services from the Ecosigma catalogue to distribute them free through their Web sites. For example, chambers include such products and services on their portals, distributing them to member companies with their own brand. Promoter organizations can contract the tools and services they are interested in for their sector with the help of Ecosigma, which facilitates their customization and integration into their portals. These products will be adapted to the promoters' brand image so that any user accessing a promoter's Web site will find all environmental adaptation products fully integrated into the network.
- **Distributors.** These are organizations that acquire a product or service from the Ecosigma catalogue in order to adapt, sell, or use it in the services they provide to their customers. This form of sale is not yet available, as all the products Ecosigma has available are free and are distributed exclusively through promoters.

Figure 2. Development of a special environmental management Web site based on Ecosigma

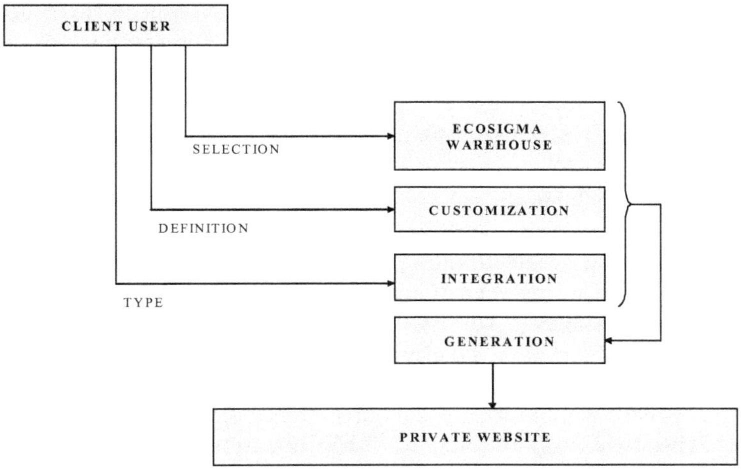

- **End users.** These are firms, which through promoters, have the opportunity to access the environmental adaptation products and services free from the platform. Each client user can eventually set up its own Web site, selecting the products and services it wants from the Ecosigma warehouse, customizing the design (Figure 2).

The platform offers a wide range of environmental adaptation products in its virtual warehouse, divided into two types: tools (computer utilities the user company will use locally, installing them on its computer, and which will help it to carry out different tasks) and services (utilities accessible and interactively provided over the Internet). The catalogue of computer services and tools integrated into the Ecosigma system has two levels: a first level includes products developed by and for the system itself; the second level covers products by outside companies that have been integrated into the network. The tools and services are all sectorial applications. An outstanding feature of the set of tools included is that the design of all the computer applications is fully integrated into the other applications and directories existing on the portal.

The Ecosigma computer platform includes a wide range of environmental management instruments. This knowledge encapsulated in the portal makes it easier for companies to access best environmental practices, draw up guides introducing environmental management systems, and free services, which allow any member fast access to what it needs to know in terms of regulations, tele-consulting, information, and European news. The system also makes a complete group of applications available to firms facilitating environmental management, among them self-diagnostic tools, the Infosigma tool, created to facilitate the processing and preservation of environmental records according to the directives of the ISO 14001 standard and the EMAS community regulations, and an administrative form manager. Finally, Ecosigma provides access to a series of advanced computerized tools, which are complementary to a company's environmental management.

Interactive system of sensors for Environmental Monitoring and Analysis (EMA). This is an electronic system for the measurement, storage, and remote monitoring of the critical environmental parameters for an industrial plant. The system allows: (1) the generation of reports from the stored measurements; (2) the configuration of the system; (3) the generation of alarms if one of the measurements exceeds the pre-programmed limit values; (4) the periodic sending of e-mails with information about the data collected; and (5) the periodic sending of messages to mobile phones with information about the data collected. EMA meets the standardization condition, which allows compatibility with the majority of sensors available on the market, as well as monitoring parameters in a wide range of industries, achieving production quality control and a reduction in pollution. In addition, the introduction of the autonomous EMA node enables the system to be fully applicable to micro-businesses and small and medium enterprises (SMEs) where it is not necessary to install a local network of nodes. EMA is integrated into the Ecosigma web server so that all users can access it remotely, either through the Internet, in a local area network (LAN), simply with a single machine, or by using a combination of the aforementioned options. The configuration of the network formed by the web server with EMA Web Services, which is in turn connected to a local area network and to the Internet, can be seen in Figure 3.

Geographical Information System (GIS). This utility has a database connected to an up-to-date Spanish road map, which allows: (1) the enlargement of the map of industrial estates with geographical locations on the GIS road map with the database of industrial enterprises recorded in Ecosigma; (2) detailed identification of the Valencian community by provinces, counties, and municipalities; and (3) displaying the map of rubbish dumps and eco-parks in the region. The GIS is an innovative utility allowing the industrial land available in the region, the exact location in each industrial estate of the firms established there, and the best route between two municipalities to be checked together with a set of applications designed to provide a service to firms wanting more information about the Valencian community industrial districts, rubbish dumps, and eco-parks and the roads that connect them.

Figure 3. Configuration of Environmental Monitoring and Analysis (EMA)

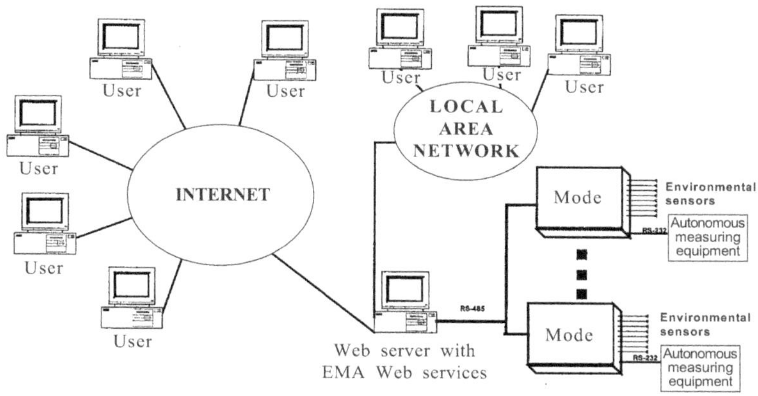

Goods Route Planning Manager. This is a service that enables companies to plan the collection of dangerous waste that they have contracted for a particular date. The module includes a complete database facilitating the management of queries about annual traffic restrictions in Spain and the European Union, an official road map for the transport of dangerous goods and a guide to good environmental practice in the transport of dangerous goods.

This modus operandi results in benefits for all agents taking part in the cooperation agreement. The improvement in profitability and efficiency in the allocation of resources has been clear, with free access to environmental management solutions with high rates of return. Value creation takes place in various ways. The integration of their products into the Ecosigma network is an attractive option for providers because it opens routes for distributing their products which they could not have contemplated before, together with a large market of potential customers free of sales costs. The self-interest of the provider companies has become an element in making the system dynamic. The system is equally attractive for the promoters as it allows them to offer their associates a range of products for very broad environmental adaptation with high added value, accessible by computer, whose development costs and complexity in terms of knowledge would make them inaccessible if the firms had decided to develop them themselves. Value creation through the network is also visible from the end users' angle. Free access to the services and tools for environmental adaptation is valuable for overcoming the asymmetries stemming from regulation, as well as for achieving the financial and strategic results the literature has associated with action that respects the natural environment.

Table 1 reports a synthesis of financial and non-financial benefits that the network has provided to the user firms. The success of the eco-efficient practices identified by Ecosigma can be seen in the eco-efficient initiatives introduced by user firms with its support, which have generated significant reductions in pollution emissions as well as in energy and other inputs consumption compatible with a financial return. The dissemination of good practices for

Table 1. Benefits provided by Ecosigma to the user firms of the network

Financial benefits				Non financial benefits	
Increased income		Cost reduction			
Ecosigma service	Benefits	Ecosigma service	Benefits	Ecosigma service	Benefits
Diffusion of good practices of environmental management. Communication plans. Sustainability Reports	Increasing sales because of growth in demand for products as an effect of increased consumer awareness of environmental risks.	Diffusion of good practices of environmental management, waste minimization techniques and cases of eco-efficiency	Reduction in energy consumption, either through more energy efficient processes or through internally generated energy making use of waste (e.g. heat) from processes.	Diffusion of good practices of environmental management and cases of eco-efficiency	Improvement in the quality of processes and products.
Diffusion of good practices of environmental management. Subproduct Auction and Exchange	Sale of sub-products resulting from the recovery or processing of waste generated in production processes.	Diffusion of good practices of environmental management, waste minimization techniques and cases of eco-efficiency	Reduction in the consumption of production factors in the processes with financially more efficient technologies.	All services.	Increase in the firm competitiveness thanks to the development of a green firm brand image.
Free access to information on aids for environmental projects	Financial aid (public or private) received for carrying out environmental investment projects.	Interactive tools for dispatch, transport and loading self-diagnosis of hazardous waste. Goods Route Planning Manager. GIS.	Reduction of waste transport, treatment and monitoring costs after its reduction and possible reuse for other purposes.	Free access to environmental legislation and information. Diffusion of good practices of environmen-tal management, and cases of eco-efficiency	Increase in knowledge stock on environmental adaptation and process technological efficiency.
		Interactive environmental management tools. Guides for emergency plans. EMA.	Reduction in costs resulting from breach of the current legislation and from pollution incidents.	Free access to Environmental Training Plans.	Increase of the environmental qualification of human capital.
				Free support to implementation of an Environ-mental Manage-ment System (EMAS or ISO 140001). EMA. Interactive tool for legal environmental self-diagnosis	Reduction in risks and the responsibilities deriving from damage caused.

Table 2. Some examples of environmental good practices, minimization techniques, and eco-efficiency cases from free access and online Ecosigma

SEARCH FOR ENVIRONMENTAL GOOD PRACTICES

Industry selection	Furniture
Process selection	Storage of inputs
	Storage and elimination of residues
	Purchases
	Stock control
	Manufacturing / water
	Manufacture / acoustic pollution
	Manufacture / emission to the atmosphere
	Manufacture / residues
	Manufacture / energetic consumption
	Formation
	Cleanliness of equipments and facilities
	Manipulation of the materials
	Preventive maintenance
	Maintenance and conservation of the facilities
	Offices and general services
Properties / advantages selection as objectives	To reduce costs
	To avoid risks and accidents
	To reduce the consumption of inputs
	To optimize the stocks of inputs
	To reduce the energetic consumption
	To diminish the water consumption
	To reduce the volume or the load pollutant of the residual waters
	To optimize the management of the atmospheric issues
	To reduce the atmospheric issues
	To reduce the acoustic pollution
	To diminish the generation of wastes
	To optimize the waste management
	To minimize the degradation of natural habitats

Results of the search		
Description	**What is obtained?**	**Advantages**
To realize a selective separation dependent on the nature of the residue	To facilitate the later management of every type of residue	To optimize the waste management
To store and to identify correctly any typology of generated residue	To control every type of generated residue	To optimize the waste management
To manage every type of residue correctly	To lead to end a suitable management of every type of generated residue	To optimize the waste management
To elaborate maps of the facilities indicating the location of containers to store each of the generated residues	To control and to facilitate the storage of the residues	To optimize the waste management
To create a common zone of storage of all the residues, identifying correctly each of them	To locate in a zone the residues generated to facilitate their management	To avoid risks and accidents. To optimize the waste management
To locate in a zone the residues generated to facilitate their management	To facilitate to fix aims of waste reduction	To optimize the waste management
To select suitable the equipments or systems of purification of the generated issues	To facilitate the direct recycling of the generated residues	To optimize the waste management
To store on cemented soil and in roofed zone the residues of secondhand{*worn*} oils. To label the above mentioned residues suitably.	To avoid possible spillages and later pollution of the soil. To control the above mentioned residues.	To avoid risks and accidents. To optimize the waste management.
To have in counts the minimization of residues in the design of new products, equipments and processes	To diminish the generation of residues in the changes of products, equipments and processes	To diminish the waste generation
To plan the production having in one counts the waste minimization	To minimize the waste generation	To diminish the waste generation. To optimize the waste management.

Table 2. continued

SEARCH FOR WASTE MINIMIZATION TECHNIQUES

Industry selection	Furniture
Process selection	Finished (application of glazes)
	Purification of atmospheric issues (organic volatile compounds)
	Purification of atmospheric issues (matter in particles)
	Water treatment (waters of painting cabin)
	Water treatment (cleanliness of glue-machines)
Properties / advantages selection as objectives	Measurements of saving of inputs
	Measurements of substitution of inputs
	Technological changes in the process (cleaner technologies)
	Organizational changes in the process or in the firm
	Improvements in the internal control and the supervision
	Measurements of energetic efficiency
	Measurements of reduction of the water consumption / production of residual waters
	Measurements of reduction of the atmospheric issues
	Measurements of reduction of the acoustic pollution
	Measurements of reduction of the waste generation
Results of the search	
Technology name	**Properties / Advantages**
Pistols of high volume and low pressure (HVLP)	- Measurements of reduction of the atmospheric issues. - Measurements of reduction of the water consumption / production of residual water. - Technological changes in the process (cleaner technologies). - Measurements of input saving.
Pulverization airless	- Measurements of input saving. - Technological changes in the process (cleaner technologies). - Measurements of reduction of the water consumption / production of residual water. - Measurements of reduction of the atmospheric issues.
Pulverization airmix	- Measurements of reduction of the atmospheric issues. - Measurements of reduction of the water consumption / production of residual water. - Technological changes in the process (cleaner technologies). - Measurements of input saving.
Electrostatic pulverization	- Measurements of input saving. - Technological changes in the process (cleaner technologies). - Measurements of reduction of the water consumption / production of residual water. - Measurements of reduction of the atmospheric issues.

Table 2. continued

SEARCH FOR ECO-EFFICIENCY CASES

Industry selection	Furniture
Good practice or minimization technique	To change metal containers of 25 liters to packages with interior bag (2 cases: pressing and nor pressing bag)
Description	The aim of the action is the reduction of the management costs of the metal containers residues by means of the substitution of packages and, later, by means of the volume reduction of the plastic dedicated to authorized agent.
Balance	The balance that one presents later has been elaborated for a company that generates 20 metal containers of 25 liters for week (44 weeks / year). The annual production of metal containers / year is 880.

Concept	Container	Package	Pressing package
Treatment cost (• / m3)	50	50	50
Annual packings	880	880	880
Units / m3	40	688	2.000
m3 de packings / year	880/40 = 22	880/688 = 1,27	880/2.000 = 0,44
Management cost (• / year)	22 x 50 = 1.100	1,27 x 50 = 63,5	0,44 x 50 = 22
Transport (• / m3)	24	24	24
Transport (• / año)	22 x 24 = 528	1,27 x 24 = 30,5	0,44 x 24 = 10,5
Total management cost (•)	1.628	94	32,5
Cost by unit (• / unit)	1,85	0,108	0,003

Conclusion	The annual saving in waste management costs obtained by the change of packings and their later pressing is notable, happening of 1.628 • in case of the metal containers to 32,5 • in case of the pressing plastic bags.

Industry selection	Furniture
Good practice or minimization technique	Recovery of dissolvers of cleanliness by means of distillation.
Description	The aim of the action is the reduction of the waste management costs, as well as the reduction of the consumption of cleanliness dissolvers.
Balance	The balance that one presents later has been elaborated for a company that consumes 25 Kg of cleanliness dissolver for week (44 weeks / year), therefore, 1.100 dissolvent kg / year.

Concept	Dissolver to distil	Recovered dissolver	Mud
Treatment cost (•/Kg)	0,39	0,39	0,42
Quantities (Kg)	25	18,75	6,25
Management cost •)	25 x 0,39 = 9,75		6,25 x 0,42 = 2,625

Saving in management (• / 25 dissolvent Kg)	9,75 – 2,625 = 7,125
Cost of dissolver as raw material (• / Kg)	1,8
Approximate investment in recovery system (•)	75
Duration of the distillation cycle (25 dissolvent Kg)	3,311
Saving in raw material (• / 25 Kg)	18,75 x 1,8 = 33,75
Saving in management (• / 25 Kg)	7,25
Total saving (• / 25 Kg)	41
Saving for distilled Kg (• / Kg)	1,66
Consumption of dissolver to cover the investment (Kg)	3,311 x 1,66 = 1.996
Investment pay-back	1 year and 10 months

Conclusion	The buy of the recovery system of dissolvers has allowed to the company to reduce the dissolver consumption, with the consequent economic saving, and to achieve a cost saving thanks to the reduction of the costs of dangerous waste management.

the technological improvement of processes compatible with improved environmental efficiency and for minimizing waste have facilitated significant cost reductions. For example, two representative eco-efficiency practices introduced in the metal and textile sectors have generated savings above the 61% and 54% respectively of annual total cost in energy, water, and other inputs consumption. Table 2 reports some examples of environmental good practices, minimization techniques, and eco-efficiency cases that user firms can study freely and online. The improvement in productivity also stems from the introduction of product innovations; the tool for analysing the product life cycle may serve as a basis for introducing environmental criteria into its design, production, or distribution that would reduce costs. The same eco-efficiency and waste minimization practices, supported by cleaner technologies and organizational changes, can reduce costs resulting from outside environmental impacts and responsibility risks, reducing the volume of pollution load in it and cutting atmospheric emissions. The guides helping with the creation of waste minimization plans and emergency plans; the environmental records manager associated with the interactive sensor system for the environmental monitoring system: the goods route planning manager; and the legal self-diagnosis make it possible to predict potential incidents which helps with the proactive development of preventive plans, which should have a positive effect on the costs of environmental damage and penalties. The optimization of waste management allowing reuse or recycling, made easier with the sub-product exchange and the online auction system, also becomes an income source. The growth of incomes can, too, originate in the sales increase made easier by the good welcome of the *green* products in the market. The user firms of the network have a last source of income growth in the easier access to financial aids for carrying out environmental investment projects. Last, the participation of a firm in Ecosigma brings some non-financial benefits, among them: an improvement of the quality of products and processes; an improvement of a *green* firm brand image; an increase of the environmental competences of the human capital; and an increase in knowledge stock on environmental adaptation and process technological efficiency.

Analysis of the Case Study: Results Concerning the Problem of the Use of IT in Developing Strategic Alliances

The failures in the environmental information market largely show themselves as a series of information asymmetries, whose effects this project can attenuate for firms integrated into the network. The uncertainty, caused by the very complexity of information and knowledge about environmental adaptation, can be reduced with the portfolio of practices, experiences, and technologies provided by the project, with already known previous results. Breaking down the value chain into processes and these into jobs and the supply of initiatives for environmental actions, whose effectiveness has been proven in each case, provides security and confidence for decision making. In addition, although appropriate solutions for some problems related to the natural environment still do not exist, the network offers the best options available on the market.

The compilation of good practices, cases of eco-efficiency, and waste minimization techniques has facilitated valuable knowledge transfer for improving environmental management and the technological efficiency of production processes, reducing the risk associated with uncertainty about the return on the investment. Alongside this, aid for the design and implementation of an environmental information system contributes towards clearing up doubts about the financial effects of environmental adaptation. The environmental responsibility of firms integrated in the network has grown with the dissemination of environmental management innovations, such as environmental management systems; training plans; communication plans; sustainability reports; hazardous waste transport guides; technical and legal environmental self-diagnosis; the development of eco-efficiency indicator systems based on an interactive manager, an environmental record manager, and administrative forms, and the geographical information system connected to the hazardous goods route planning manager. In addition, the reduction in environmental risks has been strengthened with the introduction of good maintenance and conservation practices; emergency plans and the interactive sensor system that allows the remote monitoring of the critical environmental parameters of an industrial plant; and the generation of alarms when pre-programmed value limits are exceeded.

Then, the project offers a complete catalogue of instruments to combat "Tower of Babel" syndrome (SustainAbility et al, 1993, p. 12), which hinders firms' environmental adaptation, organizing the actions, systems, and assets available coherently in order to solve the problems of protecting the natural environment. Ecosigma helps to overcome the problem of monitoring information and knowledge for environmental adaptation, selecting the relevant information and knowledge from the market (by scanning and filtering sources) and preprocessing it in order to supply it in a practical way using guides and computerized tools. The product is a set of advanced, intensive knowledge services for environmental adaptation, which the market would find difficult to generate on its own. The absence of these services from the local regional market, and the problems of internal creation because of its nature as a public good that discourages individual investment of resources in producing it, because of the risk of externality, accentuates its value for the firms integrated into the network.

Ecosigma project is a landmark example of the value creation capacity of inter-organizational strategic networks. The web of relationships allowing products or services from many content providers to be provided in a coordinated way, passed through various distributors and promoters to countless end user firms allows transaction costs to be reduced and creates sustainable competitive advantages.

Ecosigma also provides evidence that innovation multiplies when it is the result of an interactive learning process between individuals, resources, and organizations structured in networks. The synergy is so much more powerful if the network partners include those who create specific knowledge and those who provide knowledge from the specific inputs required in the local or regional sphere for environmental adaptation.

In this project, the generation of local/regional knowledge assets is fundamental. The focus on the production of territorially located environmental adaptation resources as collective goods is the sign identifying the project. The project is a collective learning model facilitating environmental innovation by enterprises in the region, encouraging the knowledge production and transfer in this field between all agents of the innovation system.

All this has only been possible with the public impulse towards collective network learning and cooperation for creating and publicizing collective goods for environmental adaptation. The public intervention has been fundamental for agreeing to the multiple coalitions of institutions involved in regional development, industrial competitiveness, R&D, and for protecting and improving the natural environment, the development of environmental adaptation products and services, and the provision of services to advanced and knowledge-intensive firms. The institutional context underlying Ecosigma has been competent in promoting collective goods for environmental adaptation, winning the trust of innovative organizations and representatives of the business fabric in a new kind of cooperative project. The financial support of regional and European institutions has been another fundamental point in the successful development of the project.

The function of the body coordinating the network is to act as an interface between the network components, maximizing their interaction based on the virtual cooperative platform. This technological link can serve as a reference model of how to stimulate environmental adaptation by incentivizing inter-organizational networks based on IT. The Ecosigma network is a flexible virtual cooperation platform providing interactions that facilitate learning and the development of knowledge for environmental adaptation more effectively than a rigid structure. Cooperative integration in a flexible network based on the Internet does not present problems of unlearning or of the integration of communities with heterogeneous practices, thereby better stimulating radical and incremental innovation. The adaptation of the environmental knowledge of the practice communities and content providers to local and regional needs allows tacit knowledge, which is difficult to apply in practice, to be transformed into explicit knowledge, which can easily be understood and used. Maximum connectivity and receptiveness, which are some features of the project, have encouraged the assimilation of competences for environmental adaptation. Virtual access to the competences of specialized forms, pre-processed by the body coordinating the network into explicit knowledge, allows these general skills to be combined with their tacit knowledge, facilitating environmental innovation.

There are basically three advantages of this organizational model for providing advanced environmental adaptation services. Firstly, it enables service inputs to be mobilized from any point on the network; to be offered at the node where the demand arises. Secondly, network integration based on IT is compatible with the maintenance of the autonomy of each agent producing and distributing services. Finally, the coordination activity provides the network with an important critical mass for the development of environmentally innovative projects normally out of the individual reach of the associated organizations.

Analysis of the structure of the virtual network and the results of the organizations involved in it provides solid empirical evidence of the positive effect of IT on the effectiveness of strategic networks of this kind, particularly for environmental adaptation. The favourable effect provided by the project does not merely stem from cooperation on a technological base, because nowadays IT is a factor common to any kind of organization.

The success of the virtual Ecosigma network depends firstly on its capacity to create value and generate competitive advantages deriving from the complementary and specific nature of the resources and capabilities it annexes. Additionally, it must be stressed that firms access to the services and instruments deposited in the network is free when they access through the promoter organizations, which involves a cost advantage difficult for competitors outside Ecosigma to reproduce. The project involves new combinations of existing and new resources, so that the integration of traditional and modern knowledge is, at the same time, developing new and valuable knowledge.

The instrument that helps with the implementation of an environmental information system (EMS), in accordance with EMAS (e.g., Management and Audit Scheme) and ISO 14001 requirements, together with the interactive environmental management tools that facilitate self-diagnosis and the design of a system of eco-efficiency indicators, enable companies to create control panels for observing their results and a better selection from the investment opportunities in the area. The improvement in the efficiency of resource management, leading to the creation of or strengthening in cost advantages of the network firms, is firstly based on the growth of knowledge on *free lunch* type environmental quality improvements (Porter & Van der Linde, 1995). The stock of knowledge on good practices, cases of eco-efficiency, and waste minimization techniques has facilitated the improvement of the technological, environmental efficiency of production processes.

The integration of a firm in this network helps to overcome restrictions on obtaining environmental competences, which could support technological and commercial differentiation; penetration into new *green* market segments; and ecologically responsible growth strategies. The improvement in the differentiation advantages of the user firms may be supported by the Ecosigma project in various ways. Firstly, through the expansion of environmentally qualified human capital, which can be based on the fostering of training plans on the topic provided by the incorporated guide, together with knowledge transfer based on the Ecosigma web manager and the tele-consultancy service. A *green* company brand image can be fostered with the creation of environmental communication plans and sustainability reports inspired by the *Sustainability Reporting Guidelines 2002* from the Global Reporting Institute (GRI), based on the interactive instruments offered by the project. In addition, horizontal differentiation may be developed by using the information flow offered by the Web site to identify *green* market segments to concentrate on. Vertical differentiation is based on introducing product innovations with environmental criteria. The product can be specialized and differentiated from similar products with the integration of the regional resources deposited in the network, accentuating its sophistication and creating barriers to entry and imitation.

Secondly, the success of this network stems from its contribution to reducing the coordination costs of the relationship and the transaction risks implicit in potential opportunist behaviours. Ecosigma network has enabled coordination costs to be reduced with a decentralized structure. The transaction risk has been minimized with the assumption by the nodal institution and the promoters/distributors of the investment cost of the price charged by the external content provider. The services obtained on the network become collective goods, free to the end user, with opportunism in the cooperation ruled out. In addition, network transaction costs are low because only a standard computer system is required for accessing the Web site. The success of the project in minimizing the problems of specialized investment with little value outside the network should be particularly highlighted, supporting the use of standard ITI which prevent firms from incurring sunk costs if they want to leave it. The risk of asset leakage—once a member has learned what it wants to it can use it against the interests of the owner enterprise—has also been successfully combated by continually reinforcing the range of unique, inimitable assets. Because of this, the cooperation option will always be preferable to purchasing on the market or internal production.

Conclusions

The life of the network studied in this work is still limited (three years), so that the evidences on the results on that to rest the conclusions are still incipient. However, analysis of the Ecosigma project and of the results of the organizations involved provides some empirical evidence of the positive impact of IT on the effectiveness of strategic networks, particularly for environmental adaptation.

Analysis of the results of the Ecosigma case study allows a series of conclusions to be reached about the success factors that enable a strategic alliance like the one described here to be started and developed. A virtual strategic network, integrated into public policy for incentivizing the provision of advanced, knowledge-intensive services for environmental adaptation will be successful when the conditions given below exist.

1. **Tangible and intangible financial effects deriving from the environmental adaptation.** The involvement of enterprises as end users of the network depends above all on their certainty about the efficiency and effectiveness of committing resources to environmental adaptation.

   The increased knowledge about environmental adaptation deposited in the network reduces the performance uncertainty associated with investment in protecting the natural environment. The flow of eco-efficient good practices identified in the project supports the thesis that a proactive environmental strategy is compatible with business profitability and competitiveness. The financial and strategic benefits achieved with environmental investments are directly linked to the increase in knowledge about environmental adaptation deposited in the network. In particular, the identification of *free lunch* environmental quality improvement projects (Porter & Van der Linde, 1995) and *win-win* environmental solutions (Florida, 1996; Walley & Whitehead, 1994; Walker, 1999) support the implementation of instruments and practices that benefit the natural environment and the efficiency and competitive position of the organization.

   The basis for improving efficiency in allocating resources and for differentiating the firm depends directly on the variety and depth of the environmental resources offered by the network.

2. **The problem of the inefficiency of the environmental market and the need for collective goods.** The network contribution to correcting the information asymmetries implicit in the environmental information market relies on it providing knowledge assets to reduce the complexity of the problem of adaptation to the natural environment.

The problem of monitoring information and knowledge for environmental adaptation can be tackled with market scanning and filtering techniques and pre-processing to facilitate assimilation by the business. A key to the success of the network is the development of innovative systems for capturing information and environmental knowledge and for filtering, evaluating, and using it. In particular, the network allows its members to access idiosyncratic, tacit, and fragmented knowledge produced in practice communities in different professional spheres, without which its compilation and coding would be inaccessible to human capital without the qualifications and professional experience normal in such communities.

The network is a model for creating advanced, knowledge-intensive services for environmental adaptation impossible for many firms to achieve because of their absence from the market and the impossibility of developing them internally. The development of new knowledge about environmental adaptation has been crucial for the success of the project. The existence of resources shared by the network members has given decisive assistance to companies getting involved as end users. The lack of a requirement for specific investment, which mitigates the problems deriving from the nature of these services as a public good, has also helped with the success of the project.

3. **Cooperation and collective learning as an environmental adaptation strategy.** Cooperation via strategic inter-organizational networks has shown itself to be an efficient and effective solution for providing advanced environmental adaptation services, which market failures make it difficult to find or develop. The case study once again supports the central idea of the theory of cooperation on the value of strategic networks as a mechanism for obtaining competences, learning, and flexible adaptation to turbulent environments.

   The wider the web of coordinated and complementary agents the greater the effectiveness and efficiency of the network are. Obtaining economies of scale inside the network does not depend on the size of the organizations involved but rather on the overall size of the network. The greater the number of interactions within the network the stronger the synergies and the process of collective learning are.

   The network efficiency in reducing transaction costs relies on the adoption of a centralized structure and a standardized infrastructure that avoids sunk costs. Opportunist behaviour is ruled out in a model of free access to environmental information and knowledge.

   New products and services must be innovative and adapted to the local or regional context. A territorially specific knowledge base must therefore be fostered within regional development policy. Environmental innovation is based on the definition of a collective learning process, with firms, institutions, and other agents related to the knowledge development and transfer located in a given territory taking part in an atmosphere of trust. The Ecosigma experience, then seems to point to the idea of the validity of the generative growth model (Maillat, 1998), which considers regions as institutions that are competent for producing and distributing knowledge assets.

The success of the project would have been impossible without a strong public impulse for the process of collective network learning. Public intervention has been fundamental for articulating the territorial organization in an inter-organizational network in which all the key agents of the innovation system are present.

4. **Collective learning in inter-organizational networks based on IT.** The effectiveness of a decentralized network like the one studied requires the nodal institution to act as an interface between the network components, multiplying the interactions so the network gains in density and transaction volume. The stronger the flow of information and knowledge, the more effective a network of this kind is, and this is made much easier with its involvement with IT. The advantage of the virtual networks is based on its capability to mobilize advanced, knowledge-intensive services towards any point in the network where demand arises. Virtual inter-organizational networks, particularly for environmental adaptation, allow collective learning models to be developed on public goods with a critical mass of organizations with no need for geographical proximity.

   The role of IT as an element determining the formation of strategic networks and their creation of value—although there is little empirical evidence for it in the previous literature (i.e., Bensaou, 1994) besides that on its effects on the market and the hierarchy—has been clearly revealed in the case study. The positive effect of IT on the results of cooperation for environmental adaptation is due to its contribution to the effectiveness of network management. Cooperation based simply on technology is not a critical factor. Although the Ecosigma ITI—the tangible dimension of the network—does not currently suffer from reproduction, this is an ephemeral advantage because it is based on assets that can easily be obtained on the market. Sustainable competitive advantages stemming from the network are extracted from the acquisition of experience in the use of IT, not on the mere possession of them. The success of the network relies on the development of management and organizational competences for the management of IT with a view towards environmental adaptation—ITMAN—the intangible dimension of the network. ITMAN based on the network can create lasting competitive advantages based on the complementary and specific nature of the capabilities they annexe.

In addition, the implementation of IT in the network supports the creation of value in as far as it contributes to the reduction of the costs of coordinating the relationship. The case studied adopts the mutual adaptation of the firms making up the network as a coordination mechanism. IT contributes to reducing coordination costs based on mutual adaptation; improving the efficiency of the lateral links that facilitate it such as decision-making processes, horizontal information, and knowledge flows; and alleviating the coordination workload, which is reduced to the less structured problems.

The effectiveness of the network will also depend on its contribution to the reduction of transaction risks implicit in potential opportunist behaviour. This contribution is maximized when the network: (1) does not oblige its members to make specialized investments with little value outside it, adopting standard ITI; (2) minimizes the information asymmetries that make the monitoring of its members' performance difficult, establishing protocols for access to the accumulated information and knowledge; and (3) it restrains the problem of the leakage of assets made available to the network, placing maximum value on maintaining the long-term relationship in order to enjoy the continuing renewal and improvement of unique and inimitable environmental adaptation services.

Another key element for the success of the network is flexible virtual cooperation. Network integration based on IT allows all members of the network to maintain their independence. In addition, the standardization of ITI makes belonging to the network flexible, preventing the enterprise incurring sunk costs if it wants to leave. The unification of technological platforms with the integration of systems between network members maximizes the efficiency of routine transactions for environmental adaptation, freeing time and resources for the creation and application of knowledge. The fewer the problems of unlearning and of change in organizational inertia the network brings the greater its flexibility will be. A network design enabling members to adopt environmental adaptation solutions and practices suitable for its own management systems minimizes these problems. The case study constitutes a model in this sense. The connectivity and receptiveness boasted by the virtual platform foster the assimilation of technological competences for environmental adaptation deposited in the network by the content providers as general knowledge and which the members can freely transform into idiosyncratic knowledge adapted to their own routines and to solving their own problems.

Nevertheless, the creation and first functioning of Ecosigma have faced relevant problems, the majority of them born of the project cooperative character. As Camisón (1993, p. 12) indicates, the cooperation is a gradual process of approximation among independent organizations that is not exempt of specific costs, which are different from the costs of transactions inherent in the market and of the costs of organization that derive from the internal growth. The most notable costs were those of functional integration among autonomous structures, accustomed to operating with different cultures, procedures, and technologies. The flexibility of the virtual network was a decisive advantage to overcome this hobble. The principal challenge was to convince to the organizations that act like promoters (substantially, Chambers of Trade and Enterprise Associations) to share investments and benefits acting jointly to develop services of environmental fit and to stop acting independently. The acceptance by the providers of the customization and integration of their tools and services into the promoters webs, reinforcing their brand image, was a necessary agreement, which time has demonstrated to be very positive on having reinforced the network decentralization and the flexibility. Another relevant problem was born of the own complexity of the network, which forced the network members to accept a common infrastructure, when they wished for a flexible and economic central node. Both the design decentralized of the network and its rest on the information technologies allowed to achieve an acceptable commitment among the members. Less pronounced were the obstacles of competitive character. The cooperation can debilitate the competitive advantages of a firm (Jarillo, 1989, p. 74) if it allows a competitor the learning of key technologies and knowledge. The structure of the network, where the promoters play a key role like interface between providers and end users, facilitated a control of the external access for the knowledge deposited for the providers in the catalogue of Ecosigma.

References

Almeida, P., & Kogut, B. (1997). The exploration of technological diversity and geographic localization of innovation. *Small Business Economics, 9*(1), 23-31.

Andersson, T., & Wolff, R. (1996). Ecology as a challenge for management research. *Scandinavian Journal of Management, 12,* 223-231.

Antonelli, C. (1998). Localized technological change, new information technology and the knowledge-based economy: The European evidence. *Journal of Evolutionary Economics, 8,* 177-198.

Antonelli, C. (1999). The evolution of the industrial organisation of the production of knowledge. *Cambridge Journal of Economics, 23,* 243-260.

Aragón-Correa, J. A. (1998). Strategic proactivity and firm approach to the natural environment. *Academy of Management Journal, 41,* 556-567.

Aragón-Correa, J. A., & Sharma, S. (2003). A contingent resource-based view of proactive corporate environmental strategy. *Academy of Management Review, 28,* 71-88.

Asheim, B. T., & Isaksen, A. (2000). Localised knowledge, interactive learning and innovation: Between regional networks and global corporations. In E. Vatne & M. Taylor (Eds.), *The networked firm in a global world. Small firms in new environments* (pp. 163-198). Aldershot: Ashgate.

Azzone, G., Bértele, U., & Noci, G. (1997). At least we are creating environmental strategies which work. *Long Range Planning, 30,* 562-571.

Ba, S., Whinston, A., & Zhang, H. (1999, May 25-26). *Small business in the digital economy: Digital company of the future.* Paper presented at the Conference "Understanding the Digital Economy. Data, Tools and Research", Department of Commerce, Washington, DC. Retrieved February 18, 2006, from http://mitpress.mit.edu

Bansal, P., & Roth, K. (2000). Why companies so green: A model of ecological responsiveness. *Academy of Management Journal, 43,* 717-736.

Barney, J. B. (1991). Firm resources and sustained competitive advantage. *Journal of Management, 17,* 99-120.

Bensaou, B. (1994). *Interorganizational cooperation: The role of information technology. An empirical comparison of US and Japanese supplier relations.* Working Paper. (94/68/TM/SM). Fontainebleau: INSEAD.

Biddle, D. (1993). Recycling for profit: The new green business frontier. *Harvard Business Review, 71*(6), 145-156.

Bowersox, D. J., Closs, D. J., & Stank, T. P. (2000). Ten mega-trends that will revolutionize supply chain logistics. *Journal of Business Logistics, 21*(2), 1-16.

Braun, P. (2002). Digital knowledge networks: Linking communities of practice with innovation. *Journal of Business Strategies, 19*(1), 43-54.

Brown, J. S., & Duguid, P. (1991). Organizational learning and communities of practice: Towards a unified view of working, learning and innovation. *Organization Science, 2,* 40-45.

Brusco, S. (1996). Sistemas globales y sistemas locales. (Global systems and local systems.) *Información Comercial Española, 754,* 63-71.

Cachon, G. P., & Fisher, M. (2000). Supply chain inventory management and the value of shared information. *Management Science, 46,* 1032-1048.

Camisón, C. (1993). Dirección de empresas en entornos globales y abiertos: Hacia la estrategia de alianzas. (Business management in global and open environments, until the alliance strategy.) *Revista Europea de Dirección y Economía de la Empresa, 2*(3), 9-37.

Camisón, C., & Lapiedra, R. (1999a). The enabling role of information technology on the emergence of new organizational forms. *Management, 2*(3), 251-261.

Camisón, C., & Lapiedra, R. (1999b). Estrategias de cooperación verticales y sistemas de informacion inter-organizativos: Análisis de la implantación del EDI en el caso de la industria europea. (Vertical strategies and inter-organizational information systems: Analysis of the EDI implementation in the case of European car industry.) *Esic. Market, 103*, 125-139.

Cash, J. I., & Konsynski, B. R. (1985). IS redraws competitive boundaries. *Harvard Business Review, 63*(2), 134-142.

Castanias, R. P., & Helfat, C. E. (1991). Managerial resources and rents. *Journal of Management, 17,* 155-171.

Chandy, R. K., & Tellis, G. J. (2000). The incumbent's curse? Incumbency, size, and radical product innovation. *Journal of Marketing, 64*(7), 1-17.

Christmann, P. (2000). Effects of 'Best Practices' of environmental management on cost advantage: The role of complementary assets. *Academy of Management Journal, 43,* 663-680.

Clemons, E. K. (1986). Information systems for sustainable competitive advantage. *Information & Management, 11,* 131-136.

Clemons, E. K. (1991). Corporate strategies for information technology: A resource-based approach. *Computer, 24*(11), 23-32.

Clemons, E. K., & Row, M. C. (1991). Sustaining IT advantage: The role of structural differences. *MIS Quarterly, 15,* 275-292.

Clemons, E. K., & Row, M. C. (1992). Information technology and industrial corporation: The changing economics of coordination and ownership. *Journal of Management Information Systems, 9*(2), 9-28.

Cothern, C. R., & Ross, N. P. (Eds.). (1994). Uncertainties in exploring the state of the environment: An overview of environmental statistics, assessment and forecasting. In *Environmental statistics, assessment and forecasting* (pp. 1-11). Florida: CRC Press.

Dwyer, F. R., Schurr, P. H., & Oh, S. (1987, April). Developing buyer seller relationships. *Journal of Marketing, 51,* 11-27.

Dyer, J. H., & Singh, H. (1998). The relational view: Cooperative strategy and sources of interorganizational competitive advantage. *Academy of Management Review, 23,* 660-679.

Ekins, P., & Newby, L. (1998). Sustainable wealth creation at the local level in an age of globalization. *Regional Studies, 32,* 863-877.

Elkington, J. (1994). Towards the sustainable corporation: Win-win-win business strategies for sustainable development. *California Management Review, 36,* 90-100.

Feeny, D., & Ives, B. (1990). In search of sustainability: Reaping long-term advantage from investments in information technology. *Journal of Management Information Systems, 7*(1), 27-46.

Florida, R. (1996). Lean and green: The move to environmentally conscious manufacturing. *California Management Review, 39,* 80-105.

Franz, D. P., & Hefter, W. H. (1999). Providing reliable information to clients: The case of environmental remediation. *International Journal of Applied Quality Management, 2,* 211-220.

Gibbs, D. C. (1996). European environment policy: The implications for local economic development. *Regional Studies, 30,* 90-92.

Gulati, R., Nohria, N., & Zaheer, A. (2000). Strategic networks. *Strategic Management Journal, 21,* 203-215.

Hamel, G. (1991). Competition for competence and inter-partner learning within international strategic alliances [Special issue]. *Strategic Management Journal, 12*, 83-103.

Hart, S. (1995). A natural-resource-based view of the firm. *Academy of Management Review, 20,* 986-1014.

Hartman, C. L., & Stafford, E. R. (1997). Green alliances: Building new business with environmental groups. *Long Range Planning, 30,* 184-196.

Hartman, H. D. (1985). *Innovation and efficiency.* New York: Pergamon Press.

Isaksen, A., & Hauge, E. (2002). *Regional clusters in Europe.* Luxemburgo: European Commission, Observatory of European SMEs.

Jarillo, J. C. (1988). On strategic networks. *Strategic Management Journal, 9,* 31-41.

Jarillo, J. C. (1989). Ventaja competitiva y ventaja cooperativa. (Competitive advantage and cooperative advantage.) *Economía Industrial, 266,* 69-75.

Jarillo, J. C. (1993). *Strategic networks: Creating the borderless organization.* London: Butterworth-Heinemann.

Jennings, P. D., & Zandbergen, P. A. (1995). Ecologically sustainable organizations: An institutional approach. *Academy of Management Review, 20,* 1015-1052.

Kalakota, R., & Robinson, M. (1999). *E-Business: Roadmap for success.* Reading, MA: Addison-Wesley.

Kalakota, R., & Whinston, A. B. (1999). *Frontiers of electronic commerce.* Reading, MA: Addison-Wesley.

Klein, B., Crawford, R. A., & Alchian, A. A. (1978). Vertical integration, appropriable rents, and the competitive contracting process. *Journal of Law and Economics, 21,* 297-326.

Kumar, K., & van Dissel, H. G. (1996). Sustainable collaboration: Managing conflict and cooperation in interorganizational systems. *MIS Quarterly, 20,* 279-300.

Lave, J., & Wenger, E. (1991). *Situated learning: Legitimate peripheral participation.* Cambridge: Cambridge University Press.

Leifer, R., Colarelli O'Connor, G., & Rice, M. (2001). Implementing radical innovation in mature firms: The role of hubs. *Academy of Management Executive, 15,* 102-113.

Lundvall, B. A., & Johnson, B. (1994). The learning economy. *Journal of Industry Studies, 1,* 23-42.

Madhok, A. (1995). Opportunism and trust in joint venture relationships: An exploratory study and a model. *Scandinavian Journal of Management, 11,* 57-74.

Maillat, D. (1998). Innovative milieux and new generations of regional policies. *Entrepreneurial & Regional Development, 10,* 1-16.

Mata, F. J., Fuerst, W. L., & Barney, J. B. (1995). Information technology and sustained competitive advantage: A resource-based analysis. *MIS Quarterly, 19,* 487-506.

McFarlan, F. W. (1984). Information technology change the way you compete. *Harvard Business Review, 22*(3), 98-103.

Milgrom, P., & Roberts, J. (1995). Complementarities and fit: Strategy, structure and organizational change in manufacturing. *Journal of Accounting and Economics, 19,* 179-208.

Millar, J., Demaid, A. M., & Quintas, P. (1997). Trans-organizational innovation: A framework for research. *Technology Analysis & Strategic Management, 9,* 399-418.

Mintzberg, H. (1979). *The structuring of organizations: A synthesis of the research.* Englewood Cliffs: Prentice-Hall.

Muller, E. (2001). *Innovation interactions between knowledge-intensive business services and small and medium sized enterprises.* New York: Verlag.

Nooteboom, B., De Jong, G., Vossen, R. W., Helper, S., & Sako, M. (2000). Network interactions and mutual dependence: A test in the car industry. *Industry and Innovation, 7,* 117-144.

Ogawa, S. (1998). Does sticky information affect the locus of innovation? Evidence from the Japanese convenience-store industry. *Research Policy, 26,* 777-790.

Oliver, C. (1997). Sustainable competitive advantage: Combining institutional and resource based views. *Strategic Management Journal, 18,* 697-713.

Orr, J. E. (1990). Sharing knowledge, celebrating identity: Community memory in a service culture. In D. Middleton & D. Edwards (Eds.), *Collective remembering* (pp. 169-189). London: Sage Publications.

Owen, C. L., & Scherer, R. F. (1993). Social responsibility and market share. *Review of Business, 15*(1), 11-17.

Perry, C., Cavaye, A., & Coote, L. (2002). Technical and social bonds within business-to-business relationships. *Journal of Business & Industrial Marketing, 17*(1), 75-86.

Porter, M. E. (2001). Strategy and the Internet. *Harvard Business Review, 79*(2), 63-78.

Porter, M. E., & Millar, V. E. (1985). How information gives you competitive advantage. *Harvard Business Review, 63*(4), 149-160.

Porter, M. E., & Van der Linde, C. (1995). Green and competitive: Ending the stalemate. *Harvard Business Review, 73*(5), 120-134.

Post, J. E. (1994). Environmental approaches and strategies: Regulation, markets, and management education. In R. Kolluru (Ed.), *Environmental strategies handbook: A guide to effective policies and practises* (pp. 11-30). New York: McGraw-Hill.

Prince, J., & Denison, R. (1994). Developing an environmental action plan for business: Environment as a standard operating practice. A case study. In R. Kolluru (Ed.), *Environmental strategies handbook: A guide to effective policies and practises* (pp. 239-258). New York: McGraw-Hill.

Pucik, V. (1988). Strategic alliances, organizational learning, and competitive advantage: The human resource management agenda. *Human Resource Management, 27,* 77-93.

Rejeski, D. R. (1994). Exploring future environmental risks. In C. R. Cothern & N. P. Ross (Eds.), *Environmental statistics, assessment and forecasting* (pp. 267-285). Florida: CRC Press.

Rivero, S. (1991). La innovación en la empresa y la utilización de recursos externos como soporte de la misma. (The innovation in the firm and the use of external resources as support of it.) *Economía Industrial, 282,* 5-14.

Rockart, J. F., & Short, J. E. (1989). IT in the 1990s: Managing organizational interdependence. *Sloan Management Review, 31*(4), 7-17.

Roome, N. (1992). Developing environmental management strategies. *Business Strategy and the Environment, 1*(1), 11-24.

Roome, N. (1994). Business Strategy, R&D management and environmental imperatives. *R&D Management, 24*(1), 65-82.

Ross, J. W., Beath, C. M., & Goodhue, D. L. (1996). Develop long-term competitiveness through IT assets. *Sloan Management Review, 38*(1), 31-42.

Roy, S., Sivakumar, K., & Wilkinson, I. F. (2004). Innovation generation in supply chain relationships: A conceptual model and research propositions. *Journal of the Academy of Marketing Science, 32,* 61-79.

Schmidheiny, S. (1992). *Changing course: A global business perspective on development and the environment.* Cambridge: MIT Press.

Schot, J., Brand, E., & Fischer, K. (1997). The greening of industry for a sustainable future: Building an international research agenda. *Business Strategy and the Environment, 6,* 153-162.

Seidl, I., Schelske, O., Joshi, J., & Jenny, M. (2003). Entrepreneurship in biodiversity conservation and regional development. *Entrepreneurship & Regional Development, 15,* 333-350.

Sharma, S. (2000). Managerial interpretations and organizational context as predictors of corporate choice of environmental strategy. *Academy of Management Journal, 43,* 681-697.

Sharma, S., & Vrendenburg, H. (1998). Proactive corporate environmental strategy and the development of competitively valuable organizational capabilities. *Strategic Management Journal, 19,* 729-753.

Shrivastava, P. (1995a). Ecocentric management for a risky society. *Academy of Management Review, 20,* 118-137.

Shrivastava, P. (1995b, Summer). Environmental technologies and competitive advantage [Special issue]. *Strategic Management Journal, 16,* 77-91.

Shrivastava, P. (1995c). The role of corporations in achieving ecological sustainability. *Academy of Management Review, 20,* 936-960.

SustainAbility, Society for the Promotion of LCA Development (SPOLD), & Business in the Environment (BiE) (1993). *The LCA sourcebook: A European business guide to life-cycle assessment.* London: SustainAbility.

Thierstein, A., & Walser, M. (1997). Sustainable regional development: The squaring of the circle or a gimmick? *Entrepreneurship & Regional Development, 9,* 159-173.

Thompson, J. D. (1967). *Organizations in action. Social science bases of administrative theory.* New York: McGraw-Hill.

Tsang, E. W. K. (2000). Transaction cost and resource-based explanations of joint ventures: A comparison and synthesis. *Organization Studies, 21,* 215-242.

Van de Linae, C. (1993). The microeconomic implications of environmental regulation: A preliminary framework. In *Environmental policies and industrial competitiveness* (pp. 69-77). Paris: OECD.

Vastag, G., Kerekes, S., & Rondinelli, D. A. (1996). Evaluation of corporate environmental management approaches: A framework and application. *International Journal of Production Economics, 43,* 193-211.

Walker, B. (1999). Maximising net benefits through biodiversity as primary land use. *Environment and Development Economics, 4,* 204-214.

Walley, N., & Whitehead, B. (1994). It's not easy being green. *Harvard Business Review, 72*(3), 46-52.

Welford, R. (1995). *Environmental strategy and sustainable development. The corporate challenge for the 21st century*. London: Routledge.

Williamson, O. E. (1975). *Markets and hierarchies: Analysis and antitrust implications. A study in the economics of internal organization.* New York: Free Press.

Williamson, O. E. (1991). Comparative economic organization: The analysis of discrete structural alternatives. *Administrative Science Quarterly, 36,* 269-296.

Young, L., & Wilkinson, I. (1997). The space between: Towards a typology of inter-firm relations. *Journal of Business-to-Business Marketing, 4*(2), 53-97.

Endnotes

1 Financial advantages can also stem from IT increasing transaction costs for the whole market. This option is an unusual one, although it can be observed in exceptional cases, for example when market mediations monopolized and abuse of the monopoly inflates transaction costs. An example of this possibility is the case of American Airlines, when it specifically raised the transaction costs of airlines setting themselves up as threatening competitors, either by increasing the cost of them connecting to its SABRE computerized booking system or by making it more difficult for them to connect.

2 A more precise formulation, developed in more detail later on, is that IT can reduce coordination costs (or at least they do not increase them unless they are being applied incorrectly). By contrast, the effect of IT on transaction risk can be positive or negative.

About the Authors

Salustiano Martínez-Fierro is a lecturer at the University of Cádiz, Spain. His research focuses on strategic alliances and entrepreneurship. He earned his PhD in management from the University of Cádiz. His articles and papers have been published by: Academy of Management Meetings; International Conference of Iberoamerican Academy of Management; European Academy of Management; Anales de Economía y Administracion de Empresas; *International Journal of Cross Cultural Management*; International Society for the Study of Work and Organizational Values (ISSWOV), AEDEM and ACEDE Conferences; and Meeting Luso-Espanholas de Gestao Científica. He has contributed to three books on strategic alliances and work values. He is a member of the entrepreneurial session of ACEDE and AEDEM, and he participates in several research groups on strategic management and human resources and creation of cultural firms. He is a member of the Regional Entrepreneurship Monitor in Andalusia (Spain), which is part of the Global Entrepreneurship Monitor (GEM) international project, developed by Babson College and the London Business School. He was a reviewer of the ACEDE Conference of 2003 and 2005, and 4th International Meetings of Iberoamerican Academy of Management.

José Aurelio Medina-Garrido is a lecturer at the University of Cádiz, Spain. His research interests include the enabling role of information systems for strategic networks and dynamic capabilities development. He was a visiting fellow at the University of Warwick (UK) and at Real Colegio Complutense at Harvard. He has contributed to several books on IT and published articles and papers in: *International Journal of Information Technology*, *Investigaciones Europeas de Dirección y Economía de la Empresa*, the International Conference on the Dynamics of Strategy (University of Surrey), European Academy of Management International Conference, and the AEDEM and ACEDE Conferences. He is a member of the Regional Entrepreneurship Monitor in Andalusia (Spain), which is part of the Global Entrepreneurship Monitor (GEM) international project, developed by Babson College and the London Business School. He participates in several research groups on dynamic capabilities and strategic change, strategic management and human resources, and creation of cultural firms. He has been a consultant for several firms.

Jose Ruiz-Navarro is a professor of management and dean of the Department of Business Administration, University of Cádiz (Spain). He has been a visiting professor at Purdue University Kranner School of Management. Ruiz-Navarro has been a consultant for the Chamber of Industry and Commerce in Cadiz, as well as for the following industries: automotive (Ford), aeronautic (Construcciones Aeronauticas, now EADS), food, and shipping. He has been CEO at ZUR (a Promotion Office of the Spanish Ministry of Industry), Gestion 1 (Consultancy), Free Zone Port of Cadiz, INI (Spanish Industrial Public Holding), and Astilleros Españoles, S.A. (Shipyard Industry). His articles and conferences have been published by: *Long Range Planning*; *Eastern Academy of Management*; International Conference on the Dynamics of Strategy (University of Surrey); EGOS (European Group of Organizational Studies); Strategic Management Society; Conference of Management and Organizational Cognition; and *Strategic Management Journal*. He is an expert evaluator of the European Commission DG XII. He was a reviewer of the International Congress of Academy of Management—Organization and Management Theory Division, 2000; Iberoamerican Academy of Management, 1999; I, II & III Workshops on Human Resources and Strategy (Spain); and *Revista Europea de Direccion y Economia de la Empresa*.

* * *

Özen Aşık earned her BA in psychology at Bogaziçi University, Istanbul (1997). After a period of professional experience in the private sector, she returned to academia and earned her MA in human resource management at Marmara University, Istanbul (2001). She is currently a PhD candidate in management at Bogaziçi University, and works as a research and teaching assistant at Yildiz Technical University. Her research interests include career management, psychological contracts, organizational justice, leadership, and strategic human resource management.

Cristina Bayona-Sáez is an associate professor of business organisation at the Department of Business Administration, Public University of Navarre. She holds a PhD in business administration from the Public University of Navarre (Spain). She teaches business economics and operation management at the degree level. Her major research interests are centered on cooperative R&D (firm's motivations for cooperative R&D, the selection of partners, the type of research, the impact of alliances on the information set). Her work has been published in *R&D Management, Research Policy, Journal of High Technology Management Research, Management Research*, and *Journal of Asset Management.*

Sebastián Bruque-Cámara is a professor of management in the Department of Business Administration, Accounting and Sociology at the University of Jaén (Spain). He currently conducts research on organisational factors affecting IT adoption and social network's effects on workers' adaptation to technological changes. He has been a visitor researcher at the University of Nijmegen (The Netherlands) and his research has appeared in the *European Journal of Information Systems, Internet Research, Journal of High Technology Management Research,* and *Strategic Management and Technology Analysis and Technovation.*

César Camisón has a PhD in economic and business sciences and is a principal professor at University Jaume I (Spain). He has been a visiting professor at the University of Texas, Universitá Commerciale Luigi Bocconi de Milán, University of Surrey, University of Vienna, and Université de Montpellier I. He has also served as director of Strategy, Knowledge Management and Organizational Learning Research Group (GRECO). His areas of specialization are strategic management and competitiveness, entrepreneurship (SME and familiar enterprises), knowledge management, strategic alliances, and industrial districts. He has participated in five European projects in ESPRIT, FORCE, and FEDER programs. He has managed and developed more than 40 European and Spanish research projects. He is the author, co-author, and coordinator of 52 books and more than 150 articles and papers to congresses. Among reviews where he has published articles are: *Environment and Planning A, Organization Studies, Management Research, International Business Review, International Journal of Technology Management, Quality Management Journal, International Journal of Quality & Reliability Management, Total Quality Management, Tourism Management, The Tourist Review,* and *International Journal of Hospitality Management.*

Ana Rosa del Águila-Obra is an associate professor of management at University of Málaga (Spain). She holds both BA and PhD degrees in business administration. Ana has authored various books, edited by RA-MA Editorial and Alfaomega Grupo Editor, and has published in different journals including *Technovation, Facilities, Journal of Global Information Technology Management, Internet Research,* and *Electronic Networking Applications and Policy*. She has published chapters in reviewed books edited by Kluwer Academic Publishers and Idea Group Publishing. Her research interests include Internet technology adoption, business-to-business electronic commerce, interorganizational trust, and supply chain management.

Guadalupe Fuentes-Lombardo is an associate professor of strategic management in the Department of Business Administration, Accounting and Sociology at the University of Jaén (Spain). Nowadays, she is finishing her PhD on internationalization process of the family business through strategic alliances. She has been a researcher in the Family Business Chair at the University of Jaén, since it was founded in 2000. She has taught several courses at postgraduate level in the Central American University of Managua (Nicaragua). Thus, she is developing several research projects about the internationalization strategy of the family business. Her publications are about distinct topics related to family businesses.

Esteban García-Canal is an associate professor of management at the University of Oviedo (Spain) and an associate member of the Instituto Empresa y Humanismo at the University of Navarra, where he also previously taught. He earned his PhD in business administration from the University of Oviedo. His research interests are focused on the confluence between organizational economics, corporate strategy, and international business. Currently, he is a member of the editorial boards of the *Journal of International Business Studies (JIBS)* and *Revista de Empresa*.

Teresa García Marco is an associate professor of business organisation at the Department of Business Administration, Public University of Navarre. She holds a PhD in business administration from the Carlos III University (Spain) and she has been a visiting scholar at the Kellogg Graduate School of Management. She teaches business economics and economics of organizations at the degree level and corporate finance in the PhD program. Professor García-Marco's research interests center on corporate governance and cooperative R&D. Her work has been published in the *Journal of Banking and Finance*, *R&D Management*, *Research Policy*, and *Journal of High Technology Management Research*.

Fátima Guadamillas Gómez is a professor of strategic management at the Department of Law and Social Sciences in Toledo, Spain. She has a degree in economics and obtained her PhD from the University of Castilla-La Mancha. She coordinates the MBA and human resources management of this university. She teaches courses at the degree level on strategic and general management and at the MBA and PhD level on knowledge management, innovation, and strategic management. Her research focuses on innovation, knowledge management strategies, and the factors related to resource sharing and knowledge transfer. Her works have been published in specialized journals and books.

Yonca Gürol is an associate professor in management in the Department of Economics and Administrative Sciences at Yildiz Technical University. She earned her BA in business administration in 1991 at Istanbul University, Department of Business Administration, Istanbul. Two years later, she received her MBA in the same department, with an emphasis on organization and business policy. She also received her PhD in the same department in 1998, on management and organization. She is currently associate professor of management in Yildiz Technical University where she teaches Introduction to Business, Management and Organization, Strategic Management, and Management of Intellectual Capital. Her research interests include intellectual capital management, strategic management, organization theory, human resource management, and international management.

Brooke C. Kelly is currently attending the University of Central Florida and is enrolled in the Industrial/Organizational Psychology Doctoral Program. She received an MA from Boston College in counseling psychology. Her research interests include shared cognition, leadership, and coaching.

Mario Javier Donate Manzanares is an associate professor at the University of Castilla-La Mancha. He teaches courses on strategic management and entrepreneurship at the Department of Law and Social Sciences in Ciudad Real, Spain. His research mainly focuses on knowledge management, technological innovation, and strategic alliances. He is currently working on his PhD, which investigates the relationship between knowledge management and technological innovation strategies and its effect on firm performance in industrial settings.

Alfonso Miguel Márquez-García is an associate professor of strategic management in the Department of Business Administration, Accounting and Sociology at the University of Jaén (Spain). He is currently finishing his PhD on cooperation relationships analyzing the role of trust according to the target where trust is placed. He has been complutense fellow at the Real Colegio Complutense at Harvard University (USA) and has had a stay at the University of Oxford (UK). His research has appeared in *Revesco, ICADE, Managing in Uncertainty,* Kluwer Academic Publishers, the Workshop on Trust Within and Between Organisations (EIASM), and Conference Proceedings of the European Academy of Management (EURAM).

José Moyano-Fuentes is a professor of management in the Department of Business Administration, Accounting and Sociology at the University of Jaén (Spain). He currently conducts research on the effect of the ownership structure on firm performance. He also has interests on the role of technology in organizations, with current emphasis involving social networks influence on employees' technology acceptance. His research has appeared in the *Journal of Management Studies, International Journal of Management Reviews,* and *Strategic Management and Technology Analysis.*

Antonio Padilla-Meléndez is an associate professor of management at the University of Málaga (Spain). He holds degrees in economics and business administration (BA in management and PhD in business administration). He is the managing director of the research team on e-business: Electronic Commerce and Teleworking, financed by the Andalusian government. Antonio has authored various books and published in different journals including: *Technovation, The International Journal of Technological Innovation and Entrepreneurship, International Journal of Business and Economics, Facilities, Journal of Global Information Technology Management, Internet Research. Electronic Networking Applications and Policy* and *International Journal of Technology Management.* He is on the editorial board of the *International Journal of E-collaboration*, and has been ad hoc reviewer for *IEEE Transactions on Professional Communication* (2004)*, Information Resources Management Journal* (2004), Academy of Management 2004 Conference, OCIS y TIM Division (2004), *IEEE Transactions on Professional Communication* (2004), and *European Journal of Operational Research* (2001). His research interests include IT adoption and usage in SME, trust in virtual teams and teleworking schemes, and techno and academic entrepreneurship.

Antonio-Rafael Ramos-Rodríguez is a lecturer at the University of Cádiz, Spain. His research interests include bibliometric analysis of entrepreneurship research. He received his PhD in management from University of Cádiz. His articles and papers have been published in *Investigaciones Europeas de Dirección y Economía de la Empresa, Strategic Management Journal*, and the AEDEM and ACEDE Conferences. He is a member of the Regional

Entrepreneurship Monitor in Andalusia (Spain), which is part of the Global Entrepreneurship Monitor (GEM) international project, developed by Babson College and the London Business School. He participates in several research groups on Dynamic Capabilities and Strategic Change and Strategic Management and Human Resources. He is a member of the entrepreneurial session of ACEDE.

Michael A. Rosen is a PhD student in the Applied Experimental and Human Factors Psychology Program at UCF. He is a graduate research assistant at the Institute for Simulation and Training where he works on several research projects including the Multidisciplinary University Research Initiative (MURI) and the Operator Performance Under Stress (OPUS) project investigating the effects of stress on team performance as well as the FuLL TiLT project developing a scenario-based training tool for leaders of multicultural teams. His research interests include individual and team decision making, human-computer interaction, and team performance.

Eduardo Salas, PhD, is currently a professor of psychology and trustee chair at the University of Central Florida where he also holds an appointment as program director for Human Systems Integration Research Department at the Institute for Simulation and Training. Previously, he was a senior research psychologist and head of the Training Technology Development Branch of the Naval Air Warfare Center Training Systems Division for 15 years. During this period, Salas served as a principal investigator for numerous R&D programs focusing on teamwork, team training, decision making under stress, and performance assessment. Salas has co-authored over 200 journal articles and book chapters and has co-edited 11 books.

Jesús David Sánchez de Pablo González del Campo is an assistant professor of general management at the University of Castilla-La Mancha. He teaches courses on economics and general management in the Department of Chemical Sciences, Ciudad Real (Spain). His research focuses on strategic alliances and cooperation between firms. He has published some articles about leadership styles and cooperation. He is currently working on his PhD, which is about firm cooperation strategies in Spanish firms.

Pablo Sánchez-Lorda is an assistant professor of management at the University of Oviedo (Spain). He received his PhD in business administration from the University of Oviedo. His research interests include mergers and acquisitions, alliances and inter-partner relationships, corporate governance and organizational economics.

Index

U

V

W

X

www.ingramcontent.com/pod-product-compliance
Lightning Source LLC
LaVergne TN
LVHW062117200926
848484LV00007B/1701